CQMechanics: A Unification of Quantum & Classical Mechanics

Quantum/Semi-Classical Entanglement
Quantum/Classical Path Integrals
Quantum/Classical Chaos

Stephen Blaha, Ph.D.

Pingree-Hill Publishing

ISBN: 978-0-9970761-8-9

Cover Credits

Cover by Stephen Blaha. Photograph of a pre-historic wheel symbolizing the unity of Quantum and Classical Mechanics through rotations.

Rev. 00/00/01 September 22, 2016

To My Wife Margaret

Some Other Books by Stephen Blaha

All the Megaverse! Starships Exploring the Endless Universes of the Cosmos using the Baryonic Force (Blaha Research, Auburn, NH, 2014)

The Algebra of Thought & Reality: The Mathematical Basis for Plato's Theory of Ideas, and Reality Extended to Include A Priori Observers and Space-Time; Second Edition (Pingree-Hill Publishing, Auburn, NH, 2009)

SuperCivilizations: Civilizations as Superorganisms (McMann-Fisher Publishing, Auburn, NH, 2010)

Universes and Megaverses: From a New Standard Model to a Physical Megaverse; The Big Bang; Our Sister Universe's Wormhole; Origin of the Cosmological Constant, Spatial Asymmetry of the Universe, and its Web of Galaxies; A Baryonic Field between Universes and Particles; Flatverse Extended Wheeler-DeWitt Equation (Blaha Research, Auburn, NH, 2014)

PHYSICS IS LOGIC PAINTED ON THE VOID: Origin of Bare Masses and The Standard Model in Logic, U(4) Origin of the Generations, Normal and Dark Baryonic Forces, Dark Matter, Dark Energy, The Big Bang, Complex General Relativity, A Megaverse of Universe Particles (Blaha Research, Auburn, NH, 2015).

PHYSICS IS LOGIC Part II: The Theory of Everything, The Megaverse Theory of Everything, U(4)⊗U(4) Grand Unified Theory (GUT), Inertial Mass = Gravitational Mass, Unified Extended Standard Model and a New Complex General Relativity with Higgs Particles, Generation Group Higgs Particles (Blaha Research, Auburn, NH, 2015).

The Origin of Higgs ("God") Particles and the Higgs Mechanism: Physics is Logic III, Beyond Higgs – A Revamped Theory With a Local Arrow of Time, The Theory of Everything Enhanced, Why Inertial Frames are Special, Universes of the Mind (Blaha Research, Auburn, NH, 2015).

The Origin of the Eight Coupling Constants of The Theory of Everything: U(8) Grand Unified Theory of Everything (GUTE), S^8 *Coupling Constant Symmetry, Space-Time Dependent Coupling Constants, Big Bang Vacuum Coupling Constants, Physics is Logic IV* (Blaha Research, Auburn, NH, 2015).

New Types of Dark Matter, Big Bang Equipartition, and A New U(4) Symmetry in the Theory of Everything: Equipartition Principle for Fermions, Matter is 83.33% Dark, Penetrating the Veil of the Big Bang, Explicit QFT Quark Confinement and Charmonium, Physics is Logic V (Blaha Research, Auburn, NH, 2015).

The Periodic Table of the 192 Quarks and Leptons in The Theory of Everything: The U(4) Layer Group, Physics is Logic VI (Blaha Research, Auburn, NH, 2016).

New Boson Quantum Field Theory, Dark Matter Dynamics, Dark Matter Fermion Layer Mixing, Genesis of Higgs Particles, New Layer Higgs Masses, Higgs Coupling Constants, Non-Abelian Higgs Gauge Fields, Physics is Logic VII (Blaha Research, Auburn, NH, 2016)

MOND Unification of the Strong Interactions and Gravitation II Quark Confinement Linked to Large-Scale Gravity Physics is Logic IX (Blaha Research, Auburn, NH, 2016)

Available on Amazon.com, Amazon.co.uk, bn.com, and other international web sites as well as at better bookstores (through Ingram Distributors).

Preface

The relation of quantum and classical phenomena has been a subject of continuing interest. Most studies approximate the quantum description of a phenomenon to obtain a classical or semi-classical approximation. This book develops a new formalism that contains both fully quantum and classical sectors, and a continuous transformation between them that provides an intermediate partly quantum – partly classical sector. This intermediate sector can play the role of a bridge between the quantum and classical descriptions of a process. Using this new formalism we consider the case of the harmonic oscillator in detail relating the quantum oscillator through the bridge to the classical oscillator. We then develop a generalization of the Feynman path integral formalism that has both a normal quantum sector and also a 'new' classical path integral sector – again with a partly quantum-partly classical intermediate sector. Our path integral generalization yields a generalization of the Schroedinger equation with both quantum and classical 'wave function' solutions. We also apply this formalism to the Fokker-Planck equation, for which it is naturally adapted. Next we apply the new formalism to quantum field theories that are known to be chaotic, and then generate a classical sector – also with chaotic behavior. We also take the standard approach to quantum entanglement and show how to extract a semi-classical entanglement as well as a classical limit without entanglement. The Boltzmann equation is easily placed within the framework of our new formalism. We solve a special case of the Vlasov equation as an example. A special relativistic Boltzmann equation is also developed within the framework of our formalism. Lastly, we develop our formalism for boson and fermion quantum field theory. We give a sensible reason why Nature must be quantum.

CONTENTS

1. Why Quantum Theory?

A question that is not often considered in these days is the reason that Nature 'chose' to be quantum rather than based on classical, deterministic mechanics. In our Theory of Everything presented in Blaha (2015c) and subsequent books in 2015 and 2016 we assumed that al Natural phenomena were ultimately based on quantum field theory.

There is a more fundamental assumption that we could posit that leads to quantum field theory and then to quantum mechanics (which is based on quantum field theory.[1]) If we assume the following postulate:

All entities in the universe are composed of discrete particles, that are integer countable, and all interactions can ultimately be reduced to the interactions of these particles.

Then, when we define field theories, they must be quantum field theories – describe particles (quanta) – and thus the field theories must be second quantized. *Particles are integer countable*[2] whether free or in perturbation theory interaction terms. And the interactions of these theories must be based on the exchange of particles although the particles have a 'cloud' of virtual particles surrounding them. The quantum mechanics of the particles constituting atoms (matter) then follows as a consequence of quantum field theory.

Classical mechanics then becomes an approximation to quantum mechanics under certain conditions that turn out to be the common conditions of everyday experience.

In basing the origin of quantum theory on the particulate nature of the entities in the universe we assume that particles exist, and can be defined, in our mostly flat space-time (which itself is generated by amalgamations of graviton particles). We also assume that the particle concept can be extended to unusual space-time coordinates such as non-static space-times. However it became apparent many years ago when accelerating coordinate systems and other non-static[3] coordinate systems were considered, that the definition of particles in quantum field theory is problematic.[4]

---

[1] Heitler (1954) shows how the Heisenberg Uncertainty Principle is a consequence of quantum field theory using an example from Quantum Electrodynamics. The Uncertainty Principle and the Correspondence Principle lead directly to quantum mechanics. Quantum mechanics is thus a consequence of quantum field theory – not an independent fundamental theory.

[2] Theories with continuous matter have not been shown to exist experimentally.

[3] A non-static coordinate system mixes space and time coordinates.

[4] S. Blaha, Il Nuovo Cimento **49 A**, 35 (1979). ___, **49 A**, 113 (1979), which appear in appendices A and B and is discussed in the following chapters, describe how to define particles in any coordinate system. The particle definition issued is discussed in papers which they reference.

Chapters 2 and 3 describe the correct definition of particles in quantum field theory. The correct definition of bosons furnishes a basis for a better definition of the Higgs Mechanism. The necessity of higher derivative theories of Gravity and the Strong Interactions[5] to obtain explicit color confinement and to reconcile gravity theory with data on the rotation of stars around galactic centers leads to an extension of the definition of particles to have principal value propagators and thus gives *particulate* action-at-a-distance.[6]

A further issue, that emerges in perturbation theory calculations in quantum field theories, leads to the introduction of a vector field as part of each propagator that eliminates the point-like nature of particle interactions in the high energy (short distance) limit in favor of 'fuzzy' interacting particles. This extension of quantum field theory is called Two-Tier quantum field theory.[7] It is required for the Theory of Everything since a conventional renormalization procedure is not known to exist – and is not likely to exist.

The combination of features that we have developed enables us to create a divergence-free[8] Theory of Everything where particles can be defined in any static or non-static coordinate system and where bosons, neglecting interactions, are stable against decay to negative energy states.[9]

The formalism that we present can be applied to quantum and classical dynamics. We define a quantum-classical formalism, that we call *QCMechanics*, that has a fully quantum sector, a classical sector, and an intermediate sector bridging the quantum and classical sectors.

We then proceed to develop the harmonic oscillator theory within this framework. Subsequently we discuss a generalized Feynman path integral formalism, a generalized Schrödinger equation, a generalized Boltzmann equation, the Fokker-Planck equation, a generalized approach to quantum and classical chaos, and to quantum entanglement as well as semi-quantum entanglement. Our formalism applies to both Quantum Field Theory and Quantum Mechanics as well as the path integrals, the Fokker-Planck equation and the Boltzmann equation.

---

[5] Unified theory: See Blaha (2016e).
[6] See appendices A and B.
[7] See Blaha (2005a).
[8] There are no divergences in perturbation theory calculations and no need for renormalization programs to remove divergences. Physical quantities do get renormalized by finite amounts.
[9] Negative energy boson states are equivalent to classical fields.

2. Boson Particle Formulation

There are two issues confronting the usual approach to the quantization of boson fields that require resolution through a 'new' quantization procedure for boson particles. One problem is the need to quantize boson fields in unconventional coordinate systems such as accelerating coordinate systems and coordinate systems defined for highly curved space-time. The other major problem is the need to quantize boson fields in such a way that bosons of negative energy have a physical interpretation.[10]

In this chapter we will define a new quantization procedure for bosons that will eliminate both of these problems. In chapter 3 we will describe the analogous quantization procedure for fermions that will enable us to create well-defined Dirac field particle states in any coordinate system. (A filled Dirac sea of negative energy fermions will exist in this formulation as it does in the conventional formulation.)

In the case of both bosons and fermions we will see that the flat space-time, static coordinate systems that are normally used will remain valid special case approximations to the new formulations of quantum field theory.

Having resolved these problems for quantum field theories of bosons and fermions we will point out in chapter 5 that 'ordinary' quantum mechanics also has a problem with quantization in unconventional coordinate systems. There are also difficulties in the transition between classical and quantum 'analogues.' For example the transition from the classical Boltzmann equation to a quantum version is uncertain.

Using a framework analogous to our 'new' approach to the quantum field theory of bosons and fermions we will establish a generalization of quantum mechanics that contains both quantum mechanics and classical mechanics, and an intermediate mixed form. This generalization supports a smooth transition between classical mechanics and quantum mechanics. With this generalization we will be able to examine the transition from quantum to classical mechanics in detail without recourse to methods such as expansions in Planck's constant $\hbar$.

2.1 Quantization of Boson Fields in Unconventional (Static and Non-Static) Coordinate Systems

The problems associated with the definition of asymptotic particle states in arbitrary coordinate systems have been pointed out by numerous authors.[11] Our 1978 paper (Appendix A)

[10] There is no Pauli Exclusion Principle for bosons. Negative energy fermions 'fill' their Dirac negative energy sea due to the limitation of fermions to one fermion per state imposed by the Pauli Exclusion Principle.

[11] See appendix A, which contains our 1978 paper Il Nuovo Cimento **49 A**, 35, for references.

resolves this problem with a consistent procedure for the local definition of asymptotic boson particle states in any coordinate system, which may or my not have a time-like Killing vector.

The general procedure is described in section 2 in Appendix A starting with eq. 6. The boson particle interpretation is described in section 4. In this chapter we wish to bring out salient details of the procedure which we will be relevant for our consideration of the generalization of quantum mechanics that we will consider later.

The first distinctive feature of this form of boson second quantization is the use of two fields to define a second quantized boson theory. The use of *two* fields enables us to define states which correspond to quantum field particles in any coordinate system. Further they give us the scope to define, not only quantum field particle states, but also classical boson field states. States, which are composites of both classical fields and quantum particles, can also be defined.

In our formulation[12] the simplest lagrangian density for a generic massless, scalar Klein-Gordon particle is:

$$\mathcal{L} = \partial\varphi_1/\partial x_\mu \partial\varphi_2/\partial x^\mu \tag{2.1}$$

with hamiltonian density

$$\mathcal{H} = \pi_1\ \pi_2\ + \partial\varphi_1/\partial x_i \partial\varphi_2/\partial x^i \tag{2.2}$$

where i labels spatial coordinates, and $\pi_1 = \partial\varphi_2/\partial t$ and $\pi_2 = \partial\varphi_1/\partial t$. Eqs. 2.1 and 2.2 are without a potential or mass term. Eq. 6, and the discussion following it, in appendix A describe the massive boson case.

The fields can be fourier expanded in terms of creation and annihilation operators:

$$\varphi_i(\mathbf{x}, t) = \int d^3k\ [a_i(k)f_k(x) + a_i^\dagger(k)f_k^*(x)] \tag{2.3}$$

for i = 1, 2 where

$$f_k(x) = e^{-ik\cdot x}/(2\omega_k(2\pi)^3)^{½}$$

with $\omega_k = |\mathbf{k}|$ in the massless case and $\omega_k = (|\mathbf{k}|^2 + m^2)^{½\,i}$ for a massive boson.

The creation and annihilation operators satisfy the commutation relations:

$$[a_i(k), a_j^\dagger(k')] = (1 - \delta_{ij})\delta^3(\mathbf{k} - \mathbf{k'}) \tag{2.4}$$
$$[a_i(k), a_j(k')] = 0$$
$$[a_i^\dagger(k), a_j^\dagger(k')] = 0$$

for i, j = 1, 2. The vacuum state |0> satisfies

$$a_1(k)|0> = a_1^\dagger(k)|0> = 0 \tag{2.5}$$
$$a_2(k)|0> \neq 0 \qquad\qquad a_2^\dagger(k)|0> \neq 0 \tag{2.6}$$

[12] In earlier books we have called this approach to second quantization *pseudoquantum field theory*.

The dual vacuum state satisfies

$$<0|a_2(k) = <0|a_2^\dagger(k) = 0 \tag{2.7}$$

$$<0|a_1(k) \neq 0 \qquad\qquad <0|a_1^\dagger(k) \neq 0 \tag{2.8}$$

Positive energy single particle *ket* states are defined using $a_2^\dagger(k)$ while negative energy ket states are defined using $a_2(k)$. Positive energy single particle *bra* states are defined using $a_1(k)$ while negative energy bra states are defined using $a_1^\dagger(k)$.

2.1.1 Transformations to Other (Possibly Non-Static) Coordinate Systems

The preceding discussion applies directly in the rectangular coordinates with which we are familiar. In eqs. 2 – 4, and their discussion, in appendix A we show that the definition of boson field orthonormal sets according to a different definition of positive frequency is related to the definition above in eq. 2.3 by a local Bogoliubov transformation. The definition of particle states and vacua are different in general. However, as eqs. 15 – 31 show, we can define the transformation to preserve the invariance of the particle number operator and thus make the theory under a different definition of positive frequency fully unitarily equivalent to the original theory. Thus the particle interpretation of states is preserved.

The general form of the Bogoliubov transformation (eq. 23) is

$$\begin{aligned} a_i(k, \lambda_1(k), \lambda_2(k)) &= B(\lambda_1(k), \lambda_2(k))a_i(k)B^{-1}(\lambda_1(k), \lambda_2(k)) \qquad (23)\\ &= \exp(i\lambda_1(k))\cosh(\lambda_2(k))a_i(k) + \exp(-i\lambda_1(k))\sinh(\lambda_2(k))a_i^\dagger(k) \end{aligned}$$

with $B(\lambda_1(k), \lambda_2(k))$ given by eq. 24 and

$$B^{-1}(\lambda_1(k), \lambda_2(k)) = B^\dagger(\lambda_1(k), \lambda_2(k)) \tag{2.9}$$

where † indicates hermitean conjugate. The text following eq. 23 provide the definition of bra and ket states, inner products, the energy-momentum tensor, equal-time commutation relations, the Green's functions, and the perturbation theory of the 'new' formalism.

Appendix A shows the general form of transformations between type '1' and type '2' creation and annihilation operators in this excerpt:[13]

[13] Excerpt used with the kind permission of Il Nuovo Cimento A.

The equal-time commutation relations, and the self-adjointness of H and φ_2 place six constraints on the constants C_{ij} and $\tilde{C}_{ij}$ in eqs. (15) and (16). After some algebra we find that we are able to express the field operators in the form

$$(40) \quad \varphi_1(x) = \int d^3k \left[\left(\frac{\cos(\theta_1 - \theta_2)}{\sin\theta_1} A_{1k} + \frac{\sin(\theta_1 - \theta_2)}{\sin\theta_1} A_{2k} \right) f_k(x) + \left(\frac{\cos(\theta_1 - \theta_2)}{\cos\theta_1} A_{1k}^\dagger - \frac{\sin(\theta_1 - \theta_2)}{\cos\theta_1} A_{2k}^\dagger \right) f_k^*(x) \right],$$

$$(41) \quad \varphi_2(x) = \int d^3k \left[(\cos\theta_2 A_{2k} + \sin\theta_2 A_{1k}) f_k(x) + (\sin\theta_2 A_{2k}^\dagger - \cos\theta_2 A_{1k}^\dagger) f_k^*(x) \right],$$

where θ_1 and θ_2 are arbitrary constants which fix the boundary conditions of the Green's functions. (They are *not* related to the Bogoliubov transformations

Thus PseudoQuantized Field Theory resolves the particle interpretation ambiguities of second quantization in non-static coordinate systems through Bogoliubov rotations.

2.1.2 Negative Energy Bosons

Traditional boson second quantization has the problem of the absence of a barrier to the decay of positive energy states to negative energy states since the Pauli Exclusion Principle does not apply to bosons. This problem has been masked ('overcome') by a clever choice of boundary conditions that are embodied in the creation/annihilation momentum space operator conditions:

$$a|0> = 0 \qquad \text{Conventional Approach} \qquad (2.10)$$
$$a^\dagger|0> \neq 0$$

In this conventional approach the creation of negative energy boson states is eliminated *ab initio* by these conditions. Yet boson quantum fields still have a conceptual physical cloud hanging over them that spin ½ fields do not. A spin ½ particle cannot transition to negative energy because there is a filled sea of negative energy particles. No additional particles can fall into the sea due to the Pauli Exclusion Principle that forbids two fermions with the same 4-momentum and quantum numbers.

In the case of scalar particles the Pauli Exclusion Principle does not apply and so, *physically*, a *filled* negative energy sea of bosons is not possible and positive energy bosons should be able to transition to negative energy states. This problem was "resolved" by the above definition of boson vacuums to exclude transitions to negative energy. But the rationale for the definition is lacking. Dirac was once asked about this issue many years ago. He said he had a solution to the problem. However he did not present it – presumably in keeping with his well-known taciturn nature. So the issue remained an open question.

In this book and earlier work[14] we showed that a more physically satisfactory method exists for avoiding the negative energy state problem. This method relies on the use of a larger Fock space in which *negative energy states (or partially negative energy states) are interpreted as states containing classical fields or a mix of classical fields and individual boson particles.* This approach resolves the negative energy boson issue and provides a common framework for boson particles and classical boson fields.

The issue of the spontaneous decay of a positive energy boson into a negative energy state still seems to exist. However all known fundamental scalar bosons are Higgs bosons that have a vacuum expectation value and a 'heavy' quantum field part of positive energy that immediately decays into other particles such as a pair of photons. The decay of a positive energy boson to a negative energy state is precluded by a separation of the formalism into separate positive energy and negative energy sectors as shown in section 4.5 below.

2.2 Classical Field States for Bosons

Classical c-number boson fields exist in our PseudoQuantum Field Theory. A classical c-number field has the form

$$\Phi(\mathbf{x}, t) = \int d^3k\ [\alpha(k)f_k(x) + \alpha^*(k)f_k{}^*(x)] \tag{2.11}$$

A corresponding classical state is a coherent state with the form

$$|\ \Phi, \Pi\rangle = C \exp\left\{\int d^3k\ [\alpha(k)a_2{}^\dagger(k) + \alpha^*(k)a_2(k)]\right\}|0\rangle \tag{2.12}$$

and correspondingly for $\Pi(x)$ where C is a normalization constant.

The defining properties of a classical field state are:

$$\begin{aligned} \varphi_1(x)|\Phi, \Pi\rangle &= \Phi(x)|\Phi, \Pi\rangle \\ \pi_1(x)|\Phi, \Pi\rangle &= \Pi(x)|\Phi, \Pi\rangle \end{aligned} \tag{2.13}$$

where $\Phi(x)$ and $\Pi(x)$ are sharp on the states and where $\varphi_i(x)$ is given by eq. 2.3.

Additional details on coherent states, which differ somewhat from conventional coherent states such as those of Kibble[15] and others, can be found in Appendix C.

2.3 The Enigma of Higgs Particles and the Higgs Mechanism

Our PseudoQuantum Field Theory is ideally suited for describing Higgs Mechanism phenomena. In our previous work on the Standard Model, and its generalization to The Extended Standard Model described in a series of books entitled *Physics is Logic*, we showed that the fermion spectrum results from Complex Special Relativity, the gauge interactions result

---

[14] See Appendix 2-A and references therein.

[15] T. W. B. Kibble, Jour. Math. Phys. **2**, 212 (1961).

from the Reality group, the fermion generations result from the Generation group, the layers of fermions result from the U(4) Layer group, and from the combination with Complex General Relativity in our Theory of Everything. Higgs particles and the Higgs Mechanism were inserted *ad hoc* to generate particle masses and symmetry breaking effects.

The apparent recent discovery of Higgs particles at CERN seems to solidify the existence of the Higgs sector of the Standard Model and of our Extended Standard Model as described in earlier volumes of *Physics is Logic*.[16]

But whence arises Higgs particles? There does not appear to be a more fundamental cause than the need for particle masses obtained through symmetry breaking. And so the Higgs sector was an expedient mechanism. With our method of avoiding divergences in perturbation theory using Two-Tier quantum field theory the need for the Higgs Mechanism appears to have disappeared with the former need for a symmetry breaking mechanism to generate particle masses. The ElectroWeak sector has no divergences in our approach and thus does not need the renormalization program previously developed that was based on symmetry breaking using the Higgs Mechanism.

In considering the Higgs Mechanism a number of peculiarities appear that diminishes its attractiveness:

1. As remarked above, it is selective in the sense that some gauge fields have associated Higgs particles and utilize the Higgs Mechanism, and some gauge fields do not have associated Higgs particles. In particular, the ElectroWeak gauge fields, the Generation group gauge fields, the Layer group gauge fields, and the complex gravitation fields have associated Higgs particles. The strong interaction (gluon) gauge fields do not.

2. The conventional Higgs potentials have a quadratic mass term of the “wrong” sign plus a quartic interaction term, which together, generate non-zero vacuum expectation values. They obviously accomplish their goal. But the source of these potentials, and why they have their form, is unknown. One suspects a fundamental principle should be operative here.

3. One can imagine creating a Higgs microscope at some super-accelerator. Using this microscope in the presence of a (classical) condensate could enable the Uncertainty Principle to be violated. This possibility, in the case of a microscope using electromagnetic fields, was the source of a heuristic argument for the need to quantize the electromagnetic field.[17]

---

[16] Blaha (2015a) and (2015b).

[17] Heitler (1954) p. 86 provides a good discussion of the need to quantize the electromagnetic field.

4. The standard formulation of the Higgs Mechanism uses classical fields under the assumption that a path integral formulation justifies their use. While this may be true, the path integral formulation relies on implicit, unstated boundary conditions that obscure the physics of the quantum field theoretic nature of the mechanism. A direct quantum field theoretic study of the Higgs Mechanism is needed and would further elucidate its character. It is possible, and it has been shown in our earlier books, that the apparently "true" mechanism described below reveals a number of important new results in a properly formulated version of the Higgs Mechanism.

2.4 "True" Origin of an Acceptable Mass Creation Mechanism

In this chapter we are using *pseudoquantization*[18] and *pseudoquantum field theory.* It combines both quantum and classical fields within the same framework. In this extended theory vacuum expectation values appear as coherent ground states that are strictly classical in nature.

This chapter is based on our 1978 paper that appeared in the peer-reviewed journal *Physical Review D*. The paper is reproduced in appendix C for the reader's convenience.

We suggest the reader skim or read the paper before proceeding. The paper also presents a new formulation of Quantum Theory that incorporates both quantum and classical mechanics within one framework that is of interest in its own right. See chapter 4 for details. Recently, experimenters have been investigating the possibility of macroscopic and other strange quantum phenomena. The new formulation is ideally suited for tracing the transition from a quantum to a classical regime. For example, it is applicable to "large n atoms" where the outermost electrons approach classical behavior with an almost continuous energy spectrum.

2.5 Higgs-Like Vacuum Expectation Value Generation of Masses

The Higgs Mechanism is based on the appearance of non-zero, c-number vacuum expectation values for Higgs fields due to potential terms directly appearing in lagrangians.

2.5.1 Pseudoquantization of Higgs Particles

We will now consider the pseudoquantization of a scalar particle using two fields in a manner shown earlier. It will become a "Higgs" particle with a non-zero vacuum expectation value.

Using the formalism described earlier we define $\varphi_1(x)$ and $\varphi_2(x)$[19] for a generic boson suppressing any internal symmetry indices for simplicity. We define a "vacuum state" containing a coherent superposition that satisfies

$$\varphi_1(x)|\Phi, \Pi\rangle = \Phi|\Phi, \Pi\rangle \tag{2.14}$$

[18] This new formalism was first described in S. Blaha, Phys. Rev. D**17**, 994 (1978).

[19] The subscripts on the fields are not gauge symmetry indices but simply identifiers distinguishing the fields from one another.

where Φ is a constant. Evaluating a fermion interaction term we find a mass term emerges[20]

$$\bar{\psi}(\varphi_1 + \varphi_2)\psi \rightarrow \bar{\psi}(\Phi + \varphi_2)\psi \quad (2.15)$$

It can also generate a mass for an interaction with a gauge field of the form

$$A^{\mu}(\varphi_1 + \varphi_2)^2 A_{\mu} \rightarrow A^{\mu}(\Phi + \varphi_2)^2 A_{\mu} \quad (2.16)$$

for ElectroWeak and other gauge fields. The φ_2 term leads to the production of Higgs particles in interactions. (The production of Higgs particles that decay into ElectroWeak gauge particles has recently been found at CERN.)

The present formalism provides a clean way to separate the vacuum expectation value of a scalar particle from its quantum field part in contrast to the conventional Higgs Mechanism where one has to separate a Higgs field into parts manually.

To obtain both the vacuum expectation value and the interaction with the quantum part of the pseudoquantum fields we choose to always specify interactions with fermions and gauge fields using $\varphi = \varphi_1 + \varphi_2$ as seen above.

It appears that our formulation of the mass generation mechanism sheds significant light on the reason for the special prominence of inertial frames. Consider massive scalars.[21] Eq. 6 in appendix A describes a massive scalar particle. If the scalar is massive, then the rest frame particle "vacuum" coherent state below yields a non-zero expectation value Φ:

$$|\Phi, \Pi\rangle = C\exp\{[(2\pi)^3 m/2]^{1/2}\Phi[a_2^{\dagger}(\mathbf{0},m) + a_2(\mathbf{0},m)]\}|0\rangle \quad (2.17)$$

where m is a generic mass. (We note that the conventional Higgs Mechanism also has mass terms.) *Thus our pseudoquantum formalism allows us to define coherent "vacuum" states that lead to particle masses and Higgs particles.*

2.6 PseudoQuantized Non-Abelian Fields

The previous sections have considered scalar boson field theory. PseudoQuantum Field Theory also applies to non-abelian fields. See appendix B, Blaha (2016e) and earlier papers by the author.[22]

---

[20] When matrix elements with a "vacuum state" are calculated.

[21] Experiments at CERN have apparently discovered a Higgs particle with a 125 GeV/c mass.

[22] S. Blaha, Phys. Rev. **D10**, 4268 (1974); Phys. Rev. **D11**, 2921 (1975).

3. Fermion Quantization

Fermion field quantization is problematic in unconventional coordinate systems such as accelerating coordinate systems and coordinate systems defined for highly curved space-time.

In this chapter we will define a PseudoQuantization procedure for fermions that supports second quantization in non-static and unusual non-rectangular coordinate systems.

Having resolved these problems for both bosons and fermions using PseudoQuantum field theory we will see in chapter 4 that 'ordinary' quantum mechanics also has a problem with quantization in unconventional coordinate systems. It also has difficulties in the transition between classical and quantum mechanics. For example the transition from the classical Boltzmann equation to a quantum version is uncertain.

Using a framework analogous to our PseudoQuantum field theory formalisms for bosons and fermions we will establish a generalization of quantum mechanics, PseudoQuantization Mechanics, which contains both quantum mechanics and classical mechanics, and intermediate mixed mechanic states. This generalization supports a smooth transition between classical mechanics and quantum mechanics. With this generalization we will be able to examine the transition from quantum to classical mechanics in detail without recourse to methods such as expansions in Planck's constant $\hbar$.

The fermion PseudoQuantization procedure is described in section 3 in Appendix A to which the reader is referred. It is similar to boson PseudoQuantization in that it requires two fermion fields for each fermion particle. Eq. 61 and the following discussion in appendix A show a simple illustrative canonical lagrangian formulation of fermion PseudoQuantization including fourier expansions, equal time commutation relations, and creation and annihilation operators. Eq. 69 shows the general form of fourier expansions while eqs. 77 and 78 show the form restricted by anti-commutation relations and adjointness of the hamiltonian to have rotations of creation and annihilation operators: b_{1k} and b_{2k}, and d_{1k} and d_{2k} in a manner similar to the analogous boson case in eqs. 40 and 41.

Thus boson and fermion PseudoQuantization field theory both have pairs of fields associated with each particle and utilize rotations to implement unitary equivalence between static and non-static coordinate systems.

4. PseudoQuantization Mechanics – Joint Quantum-Classical Mechanics Formalism

Having established the need for paired fields for bosons and fermions using PseudoQuantization field theory to achieve unitary equivalence of quantization in both static and non-static coordinate systems we now turn to the case of quantum mechanics. Here again we find there is a problem associated with the transformations between coordinate systems in certain cases. There is also the problem in determining the transition between quantum mechanical entities and their classical equivalents.

These problems are analogous to those described in previous chapters for second quantization. Since quantum mechanics is derived from quantum field theory it is reasonable to suspect that the resolution of quantum mechanics problems will ultimately be found in an analogue of PseudoQuantization which we earlier saw resolved quantum field theory problems.

This chapter will present PseudoQuantization Mechanics,[23] which contains both fully quantum, and fully classical, sectors as well as an intermediate sector that provides a transition between the quantum and classical regimes. With this formalism we can overcome coordinate system issues as well as the challenge of the correspondence between classical and quantum physics.

4.1 Coordinate Systems Problems of Quantum Mechanics

Problems exist in coordinate transformations in certain quantum mechanical situations which are 'fixed' through the use of 'recipes' that patch over the difficulties. One example is the change of coordinates in path integrals.[24] Gutzwiller (1990) points out that there is no simple rule for general canonical transformations of coordinates in path integrals. Given the central role of path integrals in quantum theory the difficulties of canonical transformations in path integrals is of concern. Other problems with canonical transformations in quantum mechanics are also discussed in Gutzwiller (1990).

We shall develop the PseudoQuantization formalism with a view towards facilitating canonical transformations in quantum mechanical studies as well as elucidating the transition from the quantum to the classical regimes.

---

[23] Much of this chapter was presented in S. Blaha, Phys. Rev. D**17**, 994 (1978) which is reprinted in appendix C. See also S. Blaha, Phys. Rev. D**10**, 4268 (July, 1974) and Phys. Rev. D**11**, 2921 (1974). See appendix D.

[24] See pp. 202-3 in Gutzwiller (1990).

4.2 PseudoQuantization Harmonic Oscillator

The harmonic oscillator plays a central role in classical and quantum mechanics due to its appearance in a variety of physical problems. In this section we will describe the PseudoQuantum formulation of the one-dimensional simple harmonic oscillator as a prelude to the general PseudoQuantum description.

Appendix C contains a paper on the PseudoQuantum harmonic oscillator.[25] In this chapter we will describe it, in part, with some changes, as a formalism that embodies both classical and quantum sectors, and provides a graceful transition between classical and quantum harmonic oscillator dynamics. Thus we will have an example of a new approach to understanding the classical-quantum transition. In later chapters we will apply this approach to physical phenomena where the transition from a classical description to a quantum equivalent is problematic.

We begin with (appendix C) two commuting variables x_1 and p_1, which we augment with two new variables x_2 and p_2, defined by

$$x_i = (m\omega/\hbar)^{-1/2}\, Q_i \qquad (4.1)$$
$$p_i = (m\omega\hbar)^{1/2}\, P_i$$

for i, j = 1, 2 where

$$P_2 = -i\, d/dQ_1 \qquad (4.2)$$
$$Q_2 = i\, d/dP_1$$

with the commutation relations:

$$[Q_i, P_j] = i(1 - \delta_{ij}) \qquad (4.3)$$

for i, j = 1, 2.

Next we define raising and lowering operators

$$a_i = 2^{-1/2}(Q_i + iP_i) \qquad (4.4)$$
$$a_i^\dagger = 2^{-1/2}(Q_i - iP_i)$$
$$Q_i = (a_i + a_i^\dagger)/\sqrt{2}$$
$$P_i = (a_i - a_i^\dagger)/(\sqrt{2}i)$$

with

$$[a_i, a_j^\dagger] = (1 - \delta_{ij}) \qquad (4.5)$$
$$[a_i, a_j] = 0$$
$$[a_i^\dagger, a_j^\dagger] = 0$$

for i, j = 1, 2.

We now define an alternate set of raising and lowering operators that will use an angle θ to provide a continuous transition from classical to quantum (and vice versa)[26]

---

[25] See its description in appendix C – section II of S. Blaha, Phys Rev **D17**, 994 (1978). Excerpts used with the kind permission of Physical Review D.

[26] This definition differs from that appearing in appendix C.

$$b_1 = Q_1\cos\theta + iP_2\sin\theta$$
$$b_2 = -Q_2\sin\theta + iP_1\cos\theta \quad (4.6)$$

$$b_1^\dagger = Q_1\cos\theta - iP_2\sin\theta$$
$$b_2^\dagger = -Q_2\sin\theta - iP_1\cos\theta \quad (4.7)$$

Their commutation relations are

$$[b_1, b_1^\dagger] = \sin(2\theta) \quad (4.8a)$$
$$[b_2, b_2^\dagger] = -\sin(2\theta)$$
$$[b_1, b_2^\dagger] = [b_2, b_1^\dagger] = 0$$
$$[b_1, b_2] = [b_1^\dagger, b_2^\dagger] = 0$$

The PseudoQuantum Hamiltonian[27] is

$$\hat{H} = p_1p_2/m + m\omega^2x_1x_2 \quad (4.8b)$$
$$= \tfrac{1}{2}\omega(\{a_1, a_2^\dagger\} + \{a_2, a_1^\dagger\})$$
$$= \omega(P_1P_2 + Q_1Q_2)$$

In terms of the original P and Q variables we find

$$Q_1 = (b_1 + b_1^\dagger)/(2\cos\theta) \quad (4.9)$$
$$Q_2 = -(b_2 + b_2^\dagger)/(2\sin\theta)$$
$$P_1 = -i(b_2 - b_2^\dagger)/(2\cos\theta)$$
$$P_2 = -i\sin(\theta)(b_1 - b_1^\dagger)/(2\sin\theta)$$

4.2.1 Dirac Metric Operator ζ Transforming From Classical to Quantum Oscillator

At this point we define 'number' states with a_2 and $a_2^\dagger$:

$$|n_+, n_-\rangle = (a_2^\dagger)^{n_+}(a_2)^{n_-}|0,0\rangle \quad (4.10)$$

where

$$\hat{H}|n_+, n_-\rangle = \omega(n_+ - n_-)|n_+, n_-\rangle \quad (4.11)$$

In view of the commutation relations we wish to transform eq. 4.10 to

$$|n_+, n_-\rangle = (b_1^\dagger)^{n_+}(b_2^\dagger)^{n_-}|0,0\rangle \quad (4.12)$$

where the vacuum in eqs. 4.10 and 4.12 will be seen to be the same:

$$a_1^\dagger|0,0\rangle = a_1|0,0\rangle = 0 \quad (4.13)$$

---

[27] Eqs. 3, 12, 21 in appendix C with ω made explicit.

$$b_1|0,0> = b_2|0,0> = 0$$

We define a 'Dirac-like' metric operator ζ. It satisfies

$$\zeta^{-1}a_2^{\dagger}\zeta = b_1^{\dagger} \qquad (4.14)$$
$$\zeta^{-1}a_2\zeta = b_2^{\dagger}$$

We provisionally define

$$\zeta = \exp(aP_1Q_1 + bP_1^{\,2} + cQ_2P_1 + dP_2P_1 + eQ_1^{\,2} + fQ_2Q_1 + gP_2Q_1) \qquad (4.15)$$

Eqs. 4.4 and 4.14 imply the values of the constants in eq. 4.15 so that

$$\zeta = \exp[(-i\cos\theta\, P_1Q_1 - (\cos\theta)/2\, P_1^{\,2} + i\sin\theta\, Q_2P_1 - \sin\theta\, P_2P_1 - (\cos\theta)/2\, Q_1^{\,2} - \sin\theta\, Q_2Q_1 + i\sin\theta\, P_2Q_1)/\surd 2] \qquad (4.16)$$
$$= \exp[(\cos\theta\,(a_1^{\dagger 2} - a_1^{\,2})/2 - (\cos\theta)/4\,(a_1 - a_1^{\dagger})^2 + (\sin\theta)/2\,(a_2 + a_2^{\dagger})(a_1 - a_1^{\dagger}) +$$
$$+ (\sin\theta)/2\,(a_2 - a_2^{\dagger})(a_1 - a_1^{\dagger}) - (\cos\theta)/4\,(a_1 + a_1^{\dagger})^2 - (\sin\theta)/2\,(a_2 + a_2^{\dagger})(a_1 + a_1^{\dagger}) +$$
$$+ (\sin\theta)/2\,(a_2 - a_2^{\dagger})(a_1 + a_1^{\dagger}))/\surd 2]$$

$$\zeta^{-1} = \exp[(\cos\theta\,(a_1^{\dagger 2} - a_1^{\,2})/2 - (\cos\theta)/4\,(a_1 - a_1^{\dagger})^2 + (\sin\theta)/2\,(a_2 + a_2^{\dagger})(a_1 - a_1^{\dagger}) + \qquad (4.17)$$
$$+ (\sin\theta)/2\,(a_2 - a_2^{\dagger})(a_1 - a_1^{\dagger}) - (\cos\theta)/4\,(a_1 + a_1^{\dagger})^2 - (\sin\theta)/2\,(a_2 + a_2^{\dagger})(a_1 + a_1^{\dagger}) +$$
$$+ (\sin\theta)/2\,(a_2 - a_2^{\dagger})(a_1 + a_1^{\dagger}))/\surd 2]$$

We note the ground state (vacuum) explicitly satisfies:

$$|0,0> = \zeta^{-1}|0,0> \qquad (4.18)$$

by eq. 4.13.

We also note

$$\zeta^{-1}[a_2, a_1^{\dagger}]\zeta = 1$$

and

$$\zeta^{-1}[a_1, a_2^{\dagger}]\zeta = 1$$

imply

$$\zeta^{-1}a_1\zeta = b_1/\sin(2\theta) \qquad (4.19)$$
$$\zeta^{-1}a_1^{\dagger}\zeta = -b_2/\sin(2\theta)$$

using eq. 4.8.

The transformed Hamiltonian H can be expressed as

$$\hat{H}_\zeta = \zeta^{-1}\hat{H}\zeta = \tfrac{1}{2}\omega(\{b_1, b_1^{\dagger}\} - \{b_2, b_2^{\dagger}\})/\sin(2\theta) \qquad (4.20)$$

4.2.2 Classical, Intermediate, and Quantum Wave Functions

The classical $(n_+, n_-)^{th}$ coordinate space wave function (eq. 28) has the form:

$$\Psi_{n_+,n_-}(x_1, p_1, x_2, p_2, \theta) = (n_+!n_-!)^{-\frac{1}{2}}\langle x_1, p_1, x_2, p_2|(a_2^\dagger)^{n_+}(a_2)^{n_-}|0, 0\rangle \tag{4.21}$$
$$= (n_+!n_-!)^{-\frac{1}{2}}\langle x_1, p_1, x_2, p_2|\zeta\zeta^{-1}(a_2^\dagger)^{n_+}\zeta\zeta^{-1}(a_2)^{n_-}\zeta\zeta^{-1}|0, 0\rangle$$
$$= (n_+!n_-!)^{-\frac{1}{2}}\langle x_1, p_1, x_2, p_2|\zeta^\dagger b_1^{\dagger n_+} b_2^{\dagger n_-}|0,0\rangle$$

using the conventional normalization of states with the form $|n\rangle = (n_-!)^{-\frac{1}{2}}a^{\dagger n}|0\rangle$.

Next we note

$$\langle x_1, p_1, x_2, p_2|\zeta^\dagger = \langle x_1, p_1, x_2, p_2| \tag{4.22}$$

similarly to eq. 4.18.

$$b_1^\dagger = Q_1\cos\theta - iP_2\sin\theta \equiv \cos\theta\, Q_1 - \sin\theta\, d/dQ_1 = \cos\theta\, \eta_1 - \sin\theta\, \partial/\partial\eta_1 \tag{4.23}$$
$$b_2^\dagger = -Q_2\sin\theta - iP_1\cos\theta \equiv -i(\sin\theta\, d/dP_1 + \cos\theta\, P_1) = -i(\sin\theta\, \partial/\partial\eta_2 + \cos\theta\, \eta_2)$$
$$= \sin\theta\, \partial/\partial\eta_3 - \cos\theta\, \eta_3$$

where $\eta_3 = i\eta_2 = iQ_2$ and $\eta_1 = Q_1$.

Eq. 4.21 can be expressed as:

$$\Psi_{n_+,n_-}(x_1, p_1, x_2, p_2, \theta) = (n_+!n_-!)^{-\frac{1}{2}}(-1)^{n_-}[\cos\theta\, \eta_1 - \sin\theta\, \partial/\partial\eta_1]^{n_+} \cdot$$
$$\cdot [\cos\theta\, \eta_3 - \sin\theta\, \partial/\partial\eta_3]^{n_-}\langle x_1, p_1, x_2, p_2|0,0\rangle \tag{4.24}$$

The determination of

$$\Psi_{0,0} \equiv \langle x_1, p_1, x_2, p_2|0,0\rangle$$

begins with noting

$$\langle x_1, p_1, x_2, p_2|b_1|0,0\rangle = 0$$

or

$$(\cos\theta\, \eta_1 + \sin\theta\, \partial/\partial\eta_1)\Psi_{0,0} = 0$$

and

$$\langle x_1, p_1, x_2, p_2|b_2|0,0\rangle = 0$$

or

$$(\cos\theta\, \eta_3 + \sin\theta\, \partial/\partial\eta_3)\Psi_{0,0} = 0$$

These conditions require

$$\Psi_{0,0} = C \exp[-\tfrac{1}{2}\cot\theta(\eta_1^2 + \eta_3^2)] \tag{4.25}$$

where the normalization $C = [m\omega\cot\theta/(i\pi\hbar)]^{\frac{1}{2}}$ is determined by

$$1 = C^2 \int dx_1 dx_2 \exp[-\tfrac{1}{2}\cot\theta(\eta_1^2 + \eta_3^2)] \tag{4.26}$$

Then eq. 4.24 becomes

$$\Psi_{n_+,n_-}(x_1, p_1, x_2, p_2, \theta) = (n_+!n_-!)^{-\frac{1}{2}}[m\omega\cot\theta/(i\pi\hbar)]^{\frac{1}{2}}(-1)^{n_-}[\cos\theta\, \eta_1 - \sin\theta\, \partial/\partial\eta_1]^{n_+} \cdot$$

$$\cdot\ [\cos\theta\ \eta_3 - \sin\theta\ \partial/\partial\eta_3]^{n-}\exp[-½\cot\theta(\eta_1^{\,2} + \eta_3^{\,2})] \qquad (4.27)$$

$$= (n_+!n_-!)^{-½}[m\omega\cot\theta/(i\pi\hbar)]^{½}(-1)^{n-}[\cos\theta\ \eta_1\ - \sin\theta\ \partial/\partial\eta_1]^{n+}\cdot$$
$$\cdot\ [i\cos\theta\ \eta_2 + i\sin\theta\ \partial/\partial\eta_2]^{n-}\exp[-½\cot\theta(\eta_1^{\,2} - \eta_2^{\,2})]$$

Note that eq. 4.27 contains a product of Hermite polynomials if $\theta = \pi/4$. It is not surprising that we obtain quantum harmonic oscillator factors in the wave function since, as eq. 4.8a shows the b operators have conventional quantum oscillator commutation relations for $\theta = \pi/4$ – thus this value of θ corresponds to the quantum case. We note that Hermite polynomials $H_n(\eta)$ are generated by

$$(\eta\ - \partial/\partial\eta)^n \exp(-½\eta^2) = \exp(-½\eta^2)H_n(\eta) \qquad (4.28)$$

We can generalize Hermite polynomials for other values of θ with

$$H_n(\eta, \theta) = \exp[+½\cot\theta\ \eta^2]\ [\cos\theta\ \eta\ - \sin\theta\ \partial/\partial\eta]^n\exp[-½\cot\theta\ \eta^2] \qquad (4.29)$$

Then eq. 4.27 can be expressed by

$$\Psi_{n+,n-}(x_1, p_1, x_2, p_2, \theta) = (n_+!n_-!)^{-½}[m\omega\cot\theta/(i\pi\hbar)]^{½}(-1)^{n-}H_{n+}(\eta_1,\theta)H_{n-}(\eta_3,\theta)\exp[-½\cot\theta(\eta_1^{\,2} + \eta_3^{\,2})] \qquad (4.30)$$

$$= (n_+!n_-!)^{-½}[m\omega\cot\theta/(i\pi\hbar)]^{½}(-1)^{n-}H_{n+}(\eta_1,\theta)H_{n-}(i\eta_2,\theta)\exp[-½\cot\theta(\eta_1^{\,2} - \eta_2^{\,2})]$$
$$= (-1)^{n-}\Psi_{n+}(\eta_1, \theta)\Psi_{n-}(\eta_3, \theta)$$
$$= (-1)^{n-}\Psi_{n+}((m\omega)^{½}\,x_1, \theta)\Psi_{n-}(i(m\omega)^{½}\,x_2, \theta)$$

where

$$\Psi_n(\eta, \theta) = = (n!)^{-½}[m\omega\cot\theta/(i\pi\hbar)]^{1/4}H_n(\eta,\theta)\ \exp[-½\cot\theta\eta^2] \qquad (4.30a)$$

At $\theta = \pi/4$ the wave function factorizes into a harmonic oscillator wave function times an inverted harmonic oscillator wave function:

$$\Psi_{n+,n-}(x_1, p_1, x_2, p_2, \theta{=}\pi/4) = (n_+!n_-!)^{-½}[m\omega/(i\pi\hbar)]^{½}(-1)^{n-}2^{-(n_+ + n_-)/2}H_{n+}(\eta_1)H_{n-}(\eta_3)\exp[-½(\eta_1^{\,2}+\eta_3^{\,2})] \qquad (4.30b)$$

$$= (-1)^{n-}2^{-(n_+ + n_-)/2}\Psi_{n+}((m\omega)^{½}\,x_1)\Psi_{n-}(i(m\omega)^{½}\,x_2)$$

where $H_n(\eta)$ is a Hermite polynomial of degree n.

For $\theta = 0$ we find the b commutation relations (eq. 8a) are zero indicating that the wave function is classical in nature. In this case, simply substituting $\theta = 0$ would cause eq. 4.30 to 'blow up.' However for certain values of n+ and n– the limit $\theta \to 0$ yields a physically interesting result – a wave function that is a delta function similar to that appearing in eq. 43 in appendix C.

Consider first the case n+ = 1 and n– = 0. Then

$$\Psi_{1,0}(x_1, p_1, x_2, p_2, \theta) = [m\omega\cot\theta/(i\pi\hbar)]^{½}[\cos\theta\ \eta_1\ - \sin\theta\ \partial/\partial\eta_1]\exp[-½\cot\theta(\eta_1^2 + \eta_3^2)]$$

As $\theta \rightarrow 0$, and for $\eta_3 = 0$, we find

$$\begin{aligned}\Psi_{1,0}(x_1, p_1, x_2, p_2, \theta\rightarrow 0) &\rightarrow [m\omega/(i\hbar)]^{½}\eta_1(\pi\sin\theta)^{-½}\exp[-½\eta_1^2/\sin\theta] \\ &\rightarrow [m\omega/(i\hbar)]^{½}\eta_1(2\pi\sin\theta)^{-½}\exp[-\eta_1^2/(2\sin\theta)] \\ &\rightarrow [m\omega/(i\hbar)]^{½}\eta_1\delta(\eta_1) = 0\end{aligned} \quad (4.31)$$

Now consider the case n+ = 0 and n– = 0:

$$\Psi_{0,0}(x_1, p_1, x_2, p_2, \theta) = [m\omega\cot\theta/(i\pi\hbar)]^{½}\exp[-½\cot\theta(\eta_1^2 + \eta_3^2)]$$

As $\theta \rightarrow 0$, and for $\eta_3 = 0$, we find

$$\begin{aligned}\Psi_{0,0}(x_1, p_1, x_2, p_2, \theta\rightarrow 0) &\rightarrow [m\omega/(i\hbar)]^{½}(\pi\sin\theta)^{-½}\exp[-½\cot\theta\eta_1^2/\sin\theta] \\ &\rightarrow [m\omega/(i\hbar)]^{½}\eta_1(2\pi\sin\theta)^{-½}\exp[-\eta_1^2/(2\sin\theta)] \\ &\rightarrow [m\omega/(i\hbar)]^{½}\delta(\eta_1) = i^{-½}\delta(x_1) \neq 0\end{aligned} \quad (4.32)$$

using

$$\delta(\eta) = \lim_{\varepsilon\rightarrow 0} (\pi\varepsilon)^{-½}\exp[-\eta^2/\varepsilon] \quad (4.33)$$

Thus the Gaussian factor combined with the preceding $(2\pi\sin\theta)^{-½}$ grows to a delta-function wave function. *Wave functions corresponding to higher values of n+ and n– go to zero in the limit $\theta \rightarrow 0$. Only $\Psi_{0,0}(x_1, p_1, x_2, p_2, \theta\rightarrow 0)$ is non-zero.*

The introduction of the time dependence and a shift of the location of the minimum of the harmonic oscillator potential to x_0 would lead to a wave function such as:

$$\Psi_{0,0}(x_1, p_1, x_2, p_2, \theta\rightarrow 0) = i^{-½}\delta(x_1 - x_0\sin(\omega t)) \quad (4.34)$$

A similar behavior may be seen in the case $\eta_1 = 0$. Then we find a wave function with a factor of $\delta(x_2)$.

Lastly, the case of $\theta \rightarrow \pi/2$ is of interest. Eq. 4.30 yields

$$\begin{aligned}\Psi_{n_+,n_-}(x_1, x_2, \theta\rightarrow\pi/2) &= (n_+!n_-!)^{-½}[m\omega\cos\theta/(i\pi\hbar)]^{½}[-\partial/\partial\eta_1]^{n_+}[\partial/\partial\eta_3]^{n_-}\exp[-½\cos\theta(\eta_1^2+\eta_3^2)]|_{\theta\rightarrow\pi/2} \\ &= 0\end{aligned} \quad (4.35)$$

Figuratively speaking, the wave function progresses from one non-zero 'classical' wave function at $\theta = 0$, to a quantum mechanical wave function at $\theta = \pi/4$, to a zero value wave function at $\theta = \pi/2$. Thus one might say "The good Lord by giving us a quantum universe put us in a position halfway between nothingness and classical mechanics." By implementing

Quantum theory we get Second Quantization of particle fields, and thereby, integer countability of particle numbers – a distinct simplification in Nature.

4.2.3 Energy Eigenvalues

From eq. 4.8b, 4.10, and 4.11 we see that eq. 4.11 for the state

$$|n_+, n_-\rangle = (a_2^\dagger)^{n_+}(a_2)^{n_-}|0, 0\rangle$$

shows the energy of the wave function (eqs. 4.27 and 4.30) to be

$$E_{n+,n-} = (n_+ - n_-)\hbar\omega = [n_+ + ½ - (n_- + ½)]\hbar\omega \qquad (4.36)$$

Eq. 4.27 satisfies the PseudoQuantized Schrödinger equation:

$$\hat{H}\Psi_n(x_1, p_1, x_2, p_2, \theta, t) = i\partial\Psi_n(x_1, p_1, x_2, p_2, \theta, t)\partial t \qquad (4.37)$$

4.3 Wave Function as a Function of Position and Momentum

We can define the wave function in terms of x_1 and p_1 with a fourier transform:

$$\Psi_{n+,n-}(x_1, p_1, \theta) = \int dx_2\, e^{-ip_1x_2}\, \Psi_{n+,n-}(x_1, x_2, \theta) \qquad (4.38)$$

$\Psi_{n+,n-}(x_1, p_1, \theta=\pi/4)$, is a wave function for the combined 'normal', and inverted, harmonic oscillators. Thus the full PseudoQuantum theory enables us to define a wave function that is a function of both position and momentum without inconsistency. We discuss this topic in more detail later when we compare it to the Wigner distribution function.

4.4 Intermediate Classical-Quantum Wave Functions

For other values of θ in eq. 4.27 and 4.30 we obtain wave functions that are intermediate between quantum and classical operator wave functions. Later we will find it of interest to trace the evolution of a classical wave function to a quantum wave function and vice versa in the general case of non-harmonic oscillator dynamics.

It is interesting to note the dependence of the energy level spacing on the angle θ. The transformed energy (eq. 4.14 implements the transformation to the b operators) has the form:

$$\hat{H}_\zeta = \zeta^{-1}\hat{H}\zeta = ½\omega(\{b_1, b_1^\dagger\} - \{b_2, b_2^\dagger\})/\sin(2\theta) \qquad (4.20)$$

If either n_+ and n_- change by one unit, then the energy changes by

$$\Delta E = ½\omega/\sin(2\theta)$$

For $\theta = \pi/4$ (the quantum case)

$$\Delta E = \tfrac{1}{2}\omega \tag{4.39}$$

For $\theta = \pi/8$ (the quantum approaching classical case)

$$\Delta E = \tfrac{1}{2}\omega/0.383 = 1.307\omega \tag{4.40}$$

As $\theta \rightarrow 0$ (the classical case)

$$\Delta E \rightarrow \infty \tag{4.41}$$

Thus 'higher' (lower) energy states beyond the $n_+ = 0$ and $n_- = 0$ state are inaccessible energy-wise. This corresponds to our above finding that only the wave function $\Psi_{0,0}$ is non-zero in the classical limit.

5. General Formalism for a PseudoQuantized System

The basic procedure of our PseudoQuantization Formalism are described in our paper Phys. Rev **D17**, 994 (1978) reprinted in appendix C.[28] The relevant excerpt is

We shall now briefly outline the procedure for embedding a classical-mechanical system in a quantum system.[6] Consider a classical Hamiltonian system with one degree of freedom, and commuting canonical variables, x_1 and p_1, which have the equations of motion

$$\dot{x}_1 = -i[x_1, \hat{H}], \quad (1)$$

$$\dot{p}_1 = -i[p_1, \hat{H}], \quad (2)$$

where defining

$$\hat{H} = -i\left(\frac{\partial H(x_1,p_1)}{\partial p_1}\frac{\partial}{\partial x_1} - \frac{\partial H(x_1,p_1)}{\partial x_1}\frac{\partial}{\partial p_1}\right) \quad (3)$$

allows us to write Hamilton's equations in commutator form. With Sudarshan[6] we define

$$x_2 = i\frac{\partial}{\partial p_1} \quad (4)$$

and

$$p_2 = -i\frac{\partial}{\partial x_1} \quad (5)$$

so that

$$[x_1, x_2] = [p_1, p_2] = 0, \quad (6)$$

$$[x_1, p_2] = [x_2, p_1] = i, \quad (7)$$

and $\hat{H}$ can now be taken to be the operator

$$\hat{H} = \frac{\partial H(x_1,p_1)}{\partial p_1}p_2 + \frac{\partial H(x_1,p_1)}{\partial x_1}x_2. \quad (8)$$

---

[28] Excerpt used with the kind permission of Physical Review D.

It is now apparent that we can take the above quantities and equations of motion to describe a quantum mechanical system with two degrees of freedom in the "coordinate" representation where the "coordinates" are (x_1, p_1) and the canonical momenta are $\Pi = (p_2, -x_2)$. As we will see below the linearity of $\hat{H}$ in the momenta is crucial for the maintenance of the classical character of x_1 and p_1, and for the observability of the phase-space trajectory. Since we choose to identify the physical observables with the commutative algebra of the coordinate operators, x_1 and p_1, we are led to impose the superselection condition that the momenta, Π, are unobservable. As a result the Hamiltonian and other generators of canonical transformations, which are all linear in the momenta, are also unobservable. However, in each case there is an associated dynamical quantity which is observable.

The required unobservability of the momenta restricts the form of the interaction between a classical-made-quantum system and an inherently quantum system to

$$H_{int} = \Phi_1 x_2 + \Phi_2 p_2 + X , \tag{9}$$

where Φ_1, Φ_2, and X are functions of x_1, p_1, and the quantum system variables. The commutation relations of these functions are also constrained[6] by the superselection rule and the commutativity of the classical variables, x_1 and p_1, and their time derivatives. In the next section we will study the simple harmonic oscillator in order to exemplify the quantum-mechanical case described above and also for direct use in the field-theoretic generalizations of subsequent sections.

Based on the above discussion we assume that we start with a conventional Hamiltonian that we express as

$$H = H(x_1, p_1) = ½p_1^2 + V(x_1) \tag{5.1}$$

Introducing x_2 and p_2, as in eqs. 4 and 5 above, we can generalize H to a PseudoQuantum Hamiltonian $\hat{H}$:

$$\hat{H} = p_1 p_2 + x_2 \partial V/\partial x_1 \tag{5.2}$$

where V is a function of x_1.

We can introduce raising and lowering operators a_i and $a_i^\dagger$ using the procedure of chapter 4. Then we can proceed as in the harmonic oscillator case to calculate wave functions. In the next chapter we apply this procedure to the Boltzmann equation, which has some similarity to the Schrödinger equation.

6. PseudoQuantization of the Boltzmann Equation

The Boltzmann equation is a classical dynamics equation that describes the dynamics of a multi-particle system with interactions. The equivalent quantum formulation is not known. However Wigner and others have proposed possible quantum equivalents that of some of the expected features of the quantum Boltzmann function. In this chapter we will follow a procedure similar to that of chapter 4 for the Vlasov approximation. For special cases of the collision term of the Boltzmann equation we will obtain quantum equivalents.

6.1 Non-Relativistic Boltzmann Equation

The non-relativistic Boltzmann equation for identical particles of one chemical species is

$$[\mathbf{p}\cdot\nabla/m + F\cdot\partial/\partial\mathbf{p}]f = -\partial f/\partial t + (\partial f/\partial t)_{coll}$$

where f(**r**, **p**, t) is Boltzmann's probability density function. It has often been remarked that this Boltzmann equation strongly resembles the Schrödinger equation.

6.2 PseudoQuantum Form of the Boltzmann Equation

We can make the case that it even more strongly resembles the PseudoQuantized Schrödinger equation (see eq. 5.2) by defining

$$-i[\mathbf{p}_1\cdot\mathbf{p}_2/m - \mathbf{x}_2\cdot F]f = -\partial f/\partial t + (\partial f/\partial t)_{coll} \tag{6.1}$$

where $\mathbf{F} = \mathbf{F}(x_1, t)$ and

$$\mathbf{p}_2 = i\nabla \tag{6.2}$$
$$x_2 = -i\partial/\partial\mathbf{p}_1$$

Comparing eq. 6.1 with chapter 4, we find we can define

$$\hat{H} = \mathbf{p}_1\cdot\mathbf{p}_2/m + x_2\cdot\partial V/\partial\mathbf{x}_1 \tag{6.3}$$

where

$$\partial V/\partial\mathbf{x}_1 = \mathbf{F}(x_1, t)$$

Given the close similarity of the Boltzmann equation and the Schrödinger equation it is sensible to treat the solution of the equation as a 'wave function' that initially represents a classical state such as the classical harmonic oscillator wave function that we saw in chapter 4. Then we will define the Boltzmann distribution in terms of the wave function solution.

The PseudoQuantized Boltzmann wave equation is

$$\hat{H}\psi = -i\partial\psi/\partial t + i(\partial\psi/\partial t)_{coll} \tag{6.4}$$

where

$$\psi = \psi(\mathbf{x}_1, \mathbf{x}_2, \theta)$$

with θ defined later in specific cases. The value of θ determines whether ψ is classical, quantum, or in an intermediate state.

In a manner somewhat analogous to that of Wigner[29] we define a Boltzmann distribution with

$$f_q(\mathbf{r}_1, \mathbf{p}_1, t, \theta) = \int d^3r_2\, \psi(\mathbf{r}_1, \mathbf{r}_2, t, \theta)\psi^\dagger(\mathbf{r}_1, \mathbf{r}_2, t, \theta) \exp(-2i\mathbf{r}_2\cdot\mathbf{p}_1/\hbar) \tag{6.5}$$

in three spatial dimensions where we use the suffix 'q' of f_q to signify the PseudoQuantum Boltzmann probability density function $f_q(\mathbf{r}, \mathbf{p}, t)$. The function f_q can be classical, quantum, or intermediate between classical and quantum depending on the value of θ. We will consider examples that illustrate the dependence of f_q on θ. Later we will also see that our definition of f_q eliminates the problems of the Wigner density function.

6.3 PseudoQuantum Form of the Vlasov Equation

The collision-less Boltzmann equation is called the Vlasov equation. It is of interest because of the difficulties associated with solving the full Boltzmann equation. Its PseudoQuantized equivalent is

$$\hat{H}\psi = -i\partial\psi/\partial t \tag{6.6}$$

This equation has 'only' the difficulty of its solution for the various forces $\mathbf{F}(x_1, t)$.

We note that the one-dimensional version of eq. 6.6 where $V = ½\, x_1^2$ is solved in chapter 4.

6.4 PseudoQuantum Vlasov Equation Solution for a Three-dimensional Harmonic Oscillator Force with θ = π/4

The choice of θ = π/4 gives 'quantum' harmonic oscillator solutions consisting of a harmonic oscillator factor and an inverted harmonic oscillator factor.

The three-dimensional harmonic oscillator PseudoQuantum 'Hamiltonian' equation is

$$(\mathbf{p}_1\cdot\mathbf{p}_2/m + \mathbf{x}_2\cdot\mathbf{x}_1)\psi = -i\partial\psi/\partial t \tag{6.7}$$

or

$$(\mathbf{p}_1\cdot\mathbf{p}_2/(2m') + \mathbf{x}_2\cdot\mathbf{x}_1)\,\psi = -i\partial\psi/\partial t \tag{6.8}$$

---

[29] E. P. Wigner, Phys. Rev. **40**, 749 (1932).

This equation is fully separable[30] for $\theta = \pi/4$ in rectangular coordinates which we label x, y, and z. The solution is a product of one-dimensional PseudoQuantum harmonic oscillator wave function factors of the form of 4.30b:

$$\psi_{n+,n-}(\mathbf{r}_1, \mathbf{r}_2, t, \pi/4) = \Psi_{nx+,nx-}(x_{1x},p_{1x},x_{2x},p_{2x},t,\theta)\ \Psi_{ny+,ny-}(x_{1y},p_{1y},x_{2y},p_{2y},t,\theta)\ \Psi_{nz+,nz-}(x_{1z},p_{1z},x_{2z},p_{2z},t,\theta) \tag{6.9}$$

$$= (-1)^{n_{x-}+n_{y-}+n_{z-}} 2^{-(n_{x+}+n_{x-}+n_{y+}+n_{y-}+n_{z+}+n_{z-})/2} \Psi_{nx+}((m\omega)^{1/2} x_1)\Psi_{nx-}(i(m\omega)^{1/2} x_2)\Psi_{ny+}((m\omega)^{1/2} y_1)\cdot$$
$$\cdot\Psi_{ny-}(i(m\omega)^{1/2} y_2)\Psi_{nz+}((m\omega)^{1/2} z_1)\Psi_{nz-}(i(m\omega)^{1/2} z_2)$$

$$= A\Psi_{nx+}((m\omega)^{1/2} x_1)\Psi_{ny+}((m\omega)^{1/2} y_1)\Psi_{nz+}((m\omega)^{1/2} z_1)\Psi_{nx-}(i(m\omega)^{1/2} x_2)\Psi_{ny-}(i(m\omega)^{1/2} y_2)\ \Psi_{nz-}(i(m\omega)^{1/2} z_2)$$
$$= A\Psi_1(\mathbf{r}_1)\Psi_2(\mathbf{r}_2) \tag{6.9a}$$

with the time dependence not displayed and where

$$A = (-1)^{n_{x-}+n_{y-}+n_{z-}} 2^{-(n_{x+}+n_{x-}+n_{y+}+n_{y-}+n_{z+}+n_{z-})/2} \tag{6.9c}$$

using eq. 4.30b.

The energy, which is constant since the Hamiltonian is not explicitly time dependent, is

$$E_{n+,n-} = (n_+ - n_-)\hbar\omega \tag{6.10}$$

with

$$n_+ = n_{x+} + n_{y+} + n_{z+} \tag{6.11}$$
$$n_- = n_{x-} + n_{y-} + n_{z-}$$

Following Wigner, a fourier transform for $\theta = \pi/4$ of eq. 6.9a factors gives a *quantum* Boltzmann density function:

$$f_q(\mathbf{q}, \mathbf{p}_1, t, \pi/4) = \int d^3Q\ \Psi(\mathbf{q} - \mathbf{Q})\Psi^\dagger(\mathbf{q} + \mathbf{Q}) \exp(-2iQ\cdot\mathbf{p}_1/\hbar)$$

$$= \int d^3Q\ \Psi_1(\mathbf{q} - \mathbf{Q})\Psi_1^\dagger(\mathbf{q} + \mathbf{Q})\Psi_2(\mathbf{q} - \mathbf{Q})\Psi_2^\dagger(\mathbf{q} + \mathbf{Q}) \exp(-2iQ\cdot\mathbf{p}_1/\hbar) \tag{6.12}$$

where we let $\mathbf{r}_1 = \mathbf{q} - \mathbf{Q}$ and $\mathbf{r}_2 = \mathbf{q} + \mathbf{Q}$.

If we define the fourier transform of a wave function $\Psi(\mathbf{r})$ by

$$\Phi(\mathbf{p}) = (2\pi\hbar)^{-3/2} \int d^3r\ \Psi(\mathbf{r}) \exp(-i\mathbf{r}\cdot\mathbf{p}/\hbar)$$

then

$$f_{qp}(\mathbf{p}_1, t, \pi/4) = \int d^3q\ f_q(\mathbf{q}, \mathbf{p}_1, t, \pi/4) = \int d^3p\ \Phi(\mathbf{p})\Phi^\dagger(\mathbf{p}) \tag{6.13}$$

---

[30] For other values of θ the solutions of the equation do not separate type '1' coordinates from type '2' coordinates. See eq. 4.30 for the general case.

yields a projection of the phase space distribution into momentum space. In addition

$$f_{qp}(\mathbf{p}_1, t, \pi/4) = \Psi(\mathbf{q})\Psi^\dagger(\mathbf{q}) \tag{6.14}$$

yields a projection of the phase space distribution into coordinate space.

Thus $f_q(\mathbf{q}, \mathbf{p}_1, t, \pi/4)$ can be interpreted as the quantum equivalent of the (classical) Boltzmann distribution. *PseudoQuantization gives us 2n variables just as there are 2n variables in phase space.*

6.5 PseudoQuantum Vlasov Equation Solution for a Three-dimensional Harmonic Oscillator Force for Arbitrary θ

The general representation of our PseudoQuantized Vlasov equation solution for any value of θ is given by eq. 4.30. The 3-dimensional Vlasov representation is

$$\begin{aligned}
\psi_{n_+,n_-}(\mathbf{r}_1, \mathbf{r}_2, t, \theta) &= \psi_{n_{x+},n_{x-}}(x_{1x},p_{1x},x_{2x},p_{2x},t,\theta)\psi_{n_{y+},n_{y-}}(x_{1y},p_{1y},x_{2y},p_{2y},t,\theta)\psi_{n_{z+},n_{z-}}(x_{1z},p_{1z},x_{2z},p_{2z},t,\theta) \\
&= (-1)^{n_{x-}+n_{y-}+n_{z-}}2^{-(n_{x+}+n_{x-}+n_{y+}+n_{y-}+n_{z+}+n_{z-})/2}\Psi_{n_{x+}}((m\omega)^{1/2} x_1, t,\theta)\Psi_{n_{x-}}(i(m\omega)^{1/2} x_2, t,\theta) \cdot \\
&\quad \cdot\Psi_{n_{y+}}((m\omega)^{1/2} y_1, t,\theta)\Psi_{n_{y-}}(i(m\omega)^{1/2} y_2,t,\theta)\Psi_{n_{z+}}((m\omega)^{1/2} z_1, t,\theta)\Psi_{n_{z-}}(i(m\omega)^{1/2} z_2, t,\theta) \\
&= A\Psi_{n_{x+}}((m\omega)^{1/2} x_1, t, \theta)\Psi_{n_{y+}}((m\omega)^{1/2} y_1, t, \theta)\Psi_{n_{z+}}((m\omega)^{1/2} z_1, t, \theta)\Psi_{n_{x-}}(i(m\omega)^{1/2} x_2, t, \theta) \cdot \\
&\quad \cdot\Psi_{n_{y-}}(i(m\omega)^{1/2} y_2, t, \theta)\, \Psi_{n_{z-}}(i(m\omega)^{1/2} z_2, t, \theta) \\
&= A\Psi_1(\mathbf{r}_1, t, \theta)\Psi_2(\mathbf{r}_2, t, \theta)
\end{aligned} \tag{6.15}$$

Following similar steps as in the previous section we find

$$\begin{aligned}
f_q(\mathbf{q}, \mathbf{p}_1, t, \theta) &= \int d^3Q\, \psi_{n_+,n_-}(\mathbf{q} - \mathbf{Q}, t, \theta)\, \psi_{n_+,n_-}^\dagger(\mathbf{q} + \mathbf{Q}, t, \theta) \exp(-2iQ\cdot\mathbf{p}_1/\hbar) \\
&= A^2 \int d^3Q\, \Psi_1(\mathbf{q} - \mathbf{Q}, t, \theta)\Psi_1^\dagger(\mathbf{q} + \mathbf{Q}, t, \theta)\Psi_2(\mathbf{q} - \mathbf{Q}, t, \theta)\Psi_2^\dagger(\mathbf{q} + \mathbf{Q}, t, \theta)\exp(-2iQ\cdot\mathbf{p}_1/\hbar)
\end{aligned} \tag{6.16}$$

where we let $\mathbf{r}_1 = \mathbf{q} - \mathbf{Q}$ and $\mathbf{r}_2 = \mathbf{q} + \mathbf{Q}$.

Following similar steps we can again obtain eqs. 6.13 and 6.14 and establish a connection between phase space, and momentum and coordinate space projections.

If we define the fourier transform of a wave function $\Psi(\mathbf{r})$ by

$$\Phi(\mathbf{p}, \theta) = (2\pi\hbar)^{-3/2} \int d^3r\, \Psi(\mathbf{r}, \theta) \exp(-i\mathbf{r}\cdot\mathbf{p}/\hbar)$$

then

$$f_{qp}(\mathbf{p}_1, t, \theta) = \int d^3q\, f_q(\mathbf{q}, \mathbf{p}_1, t, \theta) = \int d^3p\, \Phi(\mathbf{p}, \theta)\Phi^\dagger(\mathbf{p}, \theta) \tag{6.17}$$

yields a projection of the phase space distribution into momentum space. In addition

$$f_{qq}(\mathbf{p}_1, t, \theta) = \Psi(\mathbf{q}, \theta)\Psi^\dagger(\mathbf{q}, \theta) \tag{6.18}$$

yields a projection of the phase space distribution into coordinate space.

Thus $f_q(\mathbf{q}, \mathbf{p}_1, t, \theta)$ can be interpreted as the quantum equivalent of the (classical) Boltzmann distribution. PseudoQuantization gives us 2n variables just as there are 2n variables in phase space.

6.6 PseudoQuantum Vlasov Equation Solution for a Three-dimensional Harmonic Oscillator Force for θ = 0 – The Classical Case

In the $\theta = 0$ case, which is the classical mechanics limit, we find that eq. 4.34 gives a precise expression for the Vlasov Boltzmann equation solution of eq. 6.16. We note that only the 0-0 wave functions are non-zero. In one dimension we have:

$$\psi = \Psi_{0,0}(x_1, p_1, x_2, p_2, \theta\to 0) = i^{-\frac{1}{2}}\delta(x_1 - x_0\sin(\omega t)) \tag{4.34}$$

The 3-dimensional case (eq. 6.16) gives

$$\begin{aligned} f_q(\mathbf{q}, \mathbf{p}_1, t, \theta{=}0) &= \int d^3Q\, \delta^3(\mathbf{q} - \mathbf{Q} - \mathbf{x}_0\sin(\omega t))\delta^3(\mathbf{q} + \mathbf{Q} - \mathbf{x}_0\sin(\omega t))\exp(-2iQ\cdot\mathbf{p}_1/\hbar) \\ &= \delta^3(\mathbf{q} - \mathbf{x}_0\sin(\omega t)) \end{aligned} \tag{6.19}$$

a classical solution specifying the classical harmonic oscillator trajectory. We note that all other solutions (for other values of n_+ and n_-) are 'pushed' to $E = \infty$ according to eq. 4.41.

We note $f_q(\mathbf{q}, \mathbf{p}_1, t, \theta{=}0)$ is positive definite as a probability should be. The integral

$$\int d^3q\, f_q(\mathbf{q}, \mathbf{p}_1, t, \theta{=}0) = 1$$

shows the sum of the probabilities of the normalized Boltzmann distribution in coordinate space is unity.

We thus have achieve a quantum-classical Boltzmann distribution in phase space in both coordinates and momenta using PseudoQuantization where the number of phase space parameters is 2n = 6 in this case—unlike the case of the Wigner density alternative.

6.7 Comparison to the Wigner Density Function

The Wigner density function in n dimensions is defined as:

$$\Psi(p, q) = \int d^nQ\, \psi(q - Q)\, \psi^\dagger(q + Q)\, \exp(-2ipQ/\hbar) \tag{6.20}$$

where $\psi(q)$ of the wave function of the system. The interpretation of $\Psi(p, q)$ as the quantum probability in phase space corresponds to f(p, q) – the classical Boltzmann distribution. It is often interpreted in that manner.

However, several concerns are usually expressed about this interpretation:

1. Although real-valued $\Psi(p, q)$ can have a negative value making a probability interpretation problematic.
2. $\Psi(p, q)$ appears to depend on 2n values. However the wave function $\psi(q)$, upon which it is defined, only depends on n variables. Thus the domains of each function are different and $\Psi(p, q)$ can only be viewed as dependent on n variables.

Wigner attempted to overcome these objections by using the quantum mechanics density matrix $\rho(q, q')$ in an attempt to reflect the usual situation that a quantum system is in a mixed state consisting of a superposition of orthogonal states $\psi_k(q)$ with a probability of $\rho_k \geq 0$ with the $\Sigma\, \rho_k = 1$. The density matrix for this case is defined to be

$$\rho(q, q') = \Sigma\, \rho_k\, \psi_k(q)\psi_k^{\dagger}(q') \qquad (6.21)$$

Using the density matrix the Wigner distribution now is

$$\Psi(p, q) = \int d^nQ\, \rho(q - Q, q + Q) \exp(-2ipQ/\hbar) \qquad (6.22)$$

The new form of the Wigner distribution is a function of 2n variables and $\Psi(p, q)$ is positive definite. However, the density matrix (eq. 6.14) has all eigenvalues between 0 and 1 and a trace equal to one. These properties are not shared by every potential Boltzmann probability f(p, q, t). Thus the representation is limited to 'special cases.'

Our form of the *quantum* Boltzmann probability distribution is $f_q(\mathbf{r}_1, \mathbf{p}_1, t, \theta)$ which we have shown overcomes the redundancy of variables in the Wigner quantum generalization of the Boltzmann distribution and gives a sensible result in the classical limit.[31]

6.8 PseudoQuantum Form of the BGK Approximation to the Boltzmann Equation

The BKG approximation to the Boltzmann equation is

$$[\mathbf{p}\cdot\nabla/m + F\cdot\partial/\partial\mathbf{p}]f = -\partial f/\partial t + \upsilon(f_0 - f) \qquad (6.23)$$

[31] Our PseudoQuantum equivalent density has the same form as the Wigner density (eq. 6.21).

where f_0 is the local Maxwell distribution $f_0 = f_0$ ($\mathbf{r}$, $\mathbf{p}$) and υ is the molecular collision frequency. This model of the collision term due to Bhatnagar, Gross, and Crook[32] has been a much studied approximation.

Before introducing the PseudoQuantum form of the BKG approximation we use the local Maxwell-Boltzmann distribution to re-express the BKG approximation in the form

$$f_0 = n[m/(2\pi kT)]^{3/2} \exp[-m(p - p_0)^2/(2kT)] \quad (6.24)$$

where n is the particle density (assumed constant at equilibrium), k is Boltzmann's constant, T is the temperature, p_0 is the average momentum, and m is the mass of a particle. Inserting f_0 in eq. 6.23 and letting

$$f = f_0 g \quad (6.25)$$

we obtain

$$[\mathbf{p}\cdot\nabla/m + \mathbf{F}\cdot\partial/\partial\mathbf{p} - (m/kT)\mathbf{F}\cdot(\mathbf{p} - \mathbf{p}_0) + \upsilon]g = -\partial g/\partial t + \upsilon \quad (6.26)$$

The PseudoQuantized equivalent of the expanded BKG approximation (eq. 6.26) is

$$[\mathbf{p}_1\cdot\mathbf{p}_2/m - \mathbf{x}_2\cdot F(\mathbf{x}_1) + (m/kT)\mathbf{F}(\mathbf{x}_1)\cdot(\mathbf{p}_2 + i\mathbf{p}_0) + i\upsilon]g = -i\partial g/\partial t + i\upsilon \quad (6.27)$$

with the Maxwell-Boltzmann distribution term acting as a 'driving force.'

Given a force F we can proceed to PseudoQuantize using operators that are similar to those of eqs. 4.1 – 4.7 but adapted to the force and the Maxwell-Boltzmann distribution 'driving force.' We will not consider BKG examples in this book although a harmonic driving force $\mathbf{F}(\mathbf{x}_1) = -m\omega^2\mathbf{x}_1$ is an interesting case to consider.

6.9 Relativistic Boltzmann Equation

The Boltzmann equation is non-relativistic and is appropriate in systems that are at rest or moving at non-relativistic velocities. If a system is traveling at relativistic velocities then the relativistic Boltzmann equation must be used. In this section we first generalize the Boltzmann equation to its special relativistic form by making all terms covariant. Then, when we 'go to' a rest frame, the relativistic equation becomes the non-relativistic Boltzmann equation.

6.9.1 Relativistic Generalization of the Boltzmann Equation

The relativistic form of the non-relativistic Boltzmann equation of section 6.1 is

$$[\mathbf{p}^\mu\nabla_\mu/m + F^\mu\partial/\partial\mathbf{p}^\mu]f = (\partial f/\partial t)_{\text{collRelativistic}} \quad (6.28)$$

[32] P. L. Bhatnagar, E. P. Gross, and M. Crook, Phys. Rev. **94**, 511 (1954).

where we use indices to transform vectors into 4-vectors: the momentum, derivative operators and the force become Lorentz 4-vectors. The collision term must now be in a relativistic form.

6.9.2 Pseudoquantized Relativistic Boltzmann Equation

The PseudoQuantum form of the Boltzmann equation is discussed earlier:

$$-i[\mathbf{p}_1\cdot\mathbf{p}_2/m - \mathbf{x}_2\cdot F]f = -\partial f/\partial t + (\partial f/\partial t)_{coll} \quad (6.1)$$
$$\mathbf{p}_2 = i\nabla \quad (6.2)$$
$$x_2 = -i\partial/\partial\mathbf{p}_1$$

We make it relativistic in a manner similar to the approach in subsection 6.9.1. The result is the relativistic PseudoQuantum Boltzmann equation:

$$[p_1{}^{\mu}p_{2\mu} - mx_2{}^{\mu}F_{\mu}(x_1{}^{\alpha})]f = im(\partial f/\partial t)_{collRelativistic} \quad (6.29)$$

where the collision term is relativistic. This formalism, superficially, has two times. However when the wave functions are calculated only one time $x_1{}^0$ is relevant as the calculations in chapter 4 in the Vlasov approximation suggest.

6.10 Quantum and Classical Entropy

The von Neumann entropy for a system described by a density matrix ρ is defined as

$$S = -\,\mathrm{tr}[\rho\ln\rho] \quad (6.30)$$

Using eigenvectors |n> the density matrix can be expressed as

$$\rho = \sum_i \eta_i|i\rangle\langle i| \quad (6.31)$$

Then ρ can be expressed in the information theory Shannon formulation of entropy:

$$S = -\sum_i \eta_i \ln \eta_i \quad (6.32)$$

If we use the harmonic oscillator development of chapter 4 we can express the von Neumann entropy in a form which ranges from classical to quantum as a function of the angle θ. If we define the harmonic oscillator states

$$|\,n+, n-\rangle = b_1^{\dagger n_+}b_2^{\dagger n_-}|0,0\rangle$$

as in chapter 4 where

$$b_1^{\dagger} = Q_1\cos\theta - iP_2\sin\theta \quad (4.23)$$

$$b_2^\dagger = -Q_2\sin\theta - iP_1\cos\theta$$

then the density matrix is

$$\rho(\theta) = \sum_{n+,n-} |n+, n-\rangle\langle n+, n-| = \sum_{n+,n-} \eta_{n+,n-}\, b_1^\dagger(\theta)^{n+} b_2^\dagger(\theta)^{n-}|0,0\rangle\langle 0,0|b_1(\theta)^{n+} b_2(\theta)^{n-} \quad (6.33)$$

In the quantum limit where $\theta = \pi/4$ we see

$$\rho(\pi/4) = \sum_{n+,n-} (\eta_{n+,n-}/2^{n+ + n-})(Q_1 - iP_2)^{n+}(Q_2 + iP_1)^{n-}|0,0\rangle\langle 0,0|(Q_1 + iP_2)^{n+}(Q_2 - iP_1)^{n-} \quad (6.34)$$

yielding a quantum density matrix and thus a von Neumann quantum entropy.

In the classical limit where $\theta = 0$ we see

$$\rho(0) = \sum_{n+,n-} (\eta_{n+,n-}/2^{n+ + n-})Q_1^{n+}P_1^{n-}|0,0\rangle\langle 0,0|Q_1^{n+}P_1^{n-} = \rho(Q_1, P_1) \quad (6.35)$$

yielding a purely classical function of Q_1 and P_1 as the density matrix. The von Neumann entropy's classical limit in this case is the classical phase space quantity

$$\begin{aligned} S &= -\{\Sigma(\eta_{n+,n-}/2^{n+ + n-})Q_1^{2n+}P_1^{2n-}\}\ln\{\Sigma(\eta_{n+,n-}/2^{n+ + n-})Q_1^{2n+}P_1^{2n-}\} \\ &= S(Q_1, P_1) \end{aligned} \quad (6.36)$$

since Q_1 and P_1 commute.

7. PseudoQuantum Path Integral Formulation

The path integral formulation of quantum mechanics (and also of quantum field theory) plays an important role in the understanding of quantum physics. One of its major issues is the transition from a quantum mechanical framework to a classical mechanical framework. We will examine this issue from the point of view of a PseudoQuantum path integral formulation. Earlier we have seen that we can embody both quantum and classical mechanics phenomena within the PseudoQuantum framework and 'rotate' between quantum and classical mechanics solutions.

In order to establish a PseudoQuantum path integral formalism we must first generate a lagrangian from a PseudoQuantum Hamiltonian. In chapter 5 we described the general formalism for deriving a PseudoQuantum Hamiltonian from a classical Hamiltonian. We now construct the PseudoQuantum lagrangian. Starting with the equations in chapter 5:

$$x_2 = i d/dp_1$$
$$p_2 = -i d/dx_1$$

$$\hat{H}(x_1, p_1, x_2, p_2) = \partial H(x_1, p_1)/\partial p_1\, p_2 + \partial H(x_1, p_1)/\partial x_1\, x_2$$

we define the velocities[33]

$$x'_1 = \partial \hat{H}(x_1, p_1, x_2, p_2)/\partial p_1 = \partial^2 H(x_1, p_1)/\partial p_1^{\,2} p_2 + \partial^2 H(x_1, p_1)/\partial x_1 \partial p_1\, x_2 \tag{7.1}$$

$$x'_2 = \partial \hat{H}(x_1, p_1, x_2, p_2)/\partial p_2 = \partial H(x_1, p_1)/\partial p_1 \big|_{p_2 = p_1} \tag{7.2}$$

The lagrangian L is constructed in the canonical way using Legendre transformations

$$L = p_1 x'_1 + p_2\, x'_2 - \hat{H}(x_1, p_1, x_2, p_2) \tag{7.3}$$

$$= p_1[\partial^2 H(x_1, p_1)/\partial p_1^{\,2} p_2 + \partial^2 H(x_1, p_1)/\partial x_1 \partial p_1\, x_2] - \partial H(x_1, p_1)/\partial x_1\, x_2$$

$$= \partial^2 H(x_1, p_1)/\partial p_1^{\,2}\, p_1 p_2 + \partial^2 H(x_1, p_1)/\partial x_1 \partial p_1\, p_1 x_2 - \partial H(x_1, p_1)/\partial x_1\, x_2 \tag{7.3a}$$

where p_1 and p_2 are extracted from eqs. 7.1 and 7.2.

We now consider the example of a harmonic oscillator where

$$H = p^2/(2m) + ½\, m\omega^2 x^2$$

---

[33] The velocity x'_2 is a defined quantity, which is defined in a manner consistent with the definition of x'_1.

$$\hat{H} = p_1p_2/m + m\omega^2x_1x_2 \qquad (4.8b)$$

Substituting in eq. 7.3 we find

$$\begin{aligned} L &= \partial^2H(x_1, p_1)/\partial p_1^2\, p_1p_2 + \partial^2H(x_1, p_1)/\partial x_1\partial p_1\, p_1x_2 - \partial H(x_1, p_1)/\partial x_1\, x_2 \\ &= p_1p_2/m - m\omega^2x_1x_2 \\ &= mx'_1x'_2 - m\omega^2x_1x_2 \end{aligned} \qquad (7.4)$$

with p_1 and p_2 determined, and replaced, as functions of x'_1 and x'_2 by eqs. 7.1 and 7.2.

The Lagrange equations of motion determined for i = 1, 2 are:

$$d/dt\,(\partial L/\partial x'_i) - \partial L/\partial x_i = 0 \qquad (7.5)$$

$$mx''_2 + m\omega^2x_2 = 0 \qquad (7.6)$$
$$mx''_1 + m\omega^2x_1 = 0 \qquad (7.7)$$

7.1 Feynman Path Integral formulation

The propagator K(x – y, t) for the Feynman path integral formulation has the form:

$$K(x - y, T) = A \lim_{n\to\infty} \int\!\!\int\!\!\int\!\!\int_{-\infty}^{+\infty} \ldots \int dx_0dx_1\ldots dx_n \exp[i/\hbar \int_t^{t+T} L(x, v, t_a)\, dt_a] \qquad (7.8)$$

where A is a constant, and the integral over the dx's ranges from $-\infty$ to ∞.

7.1.1 Conventional Formulation – Free Particle Case

In the one dimensional free particle case the path integral is a product of n infinitesimal paths of time interval ε:

$$K(x - y, T) = \prod_n G_\varepsilon \qquad (7.9)$$

where T = nδ. Using ~ to denote proportionality up to a constant we find the fourier transform of an interval of path

$$\begin{aligned} G_\delta &= \int dx\, e^{-ipx} \exp[-ix^2/(2\varepsilon)] \\ &\sim \exp[-p^2/(2\varepsilon)] \end{aligned} \qquad (7.10)$$

Then the product of the incremental factors that total to time T give

$$K(p, T) \sim \exp[-iTp^2/2] \qquad (7.11)$$

A fourier transformation yields the free particle propagator

$$\begin{aligned} K(x - y, T) &\sim \int dp\, e^{-ip(x-y)} \exp[-iTp^2/2] \\ &\sim \exp[-i(x - y)^2/T] \end{aligned} \qquad (7.12)$$

where we normalize the propagator to unity

$$\int dy\, K(x - y, T) = 1 \tag{7.13}$$

7.1.2 PeseudoQuantum Formulation – Free Particle Case

We will now develop the PseudoQuantum path integral formalism for the case of a free particle. The form of the path integral now is

$$K(x - y, T) = A \lim_{n \to \infty} \iiiint \ldots \int_{-\infty}^{+\infty} dx_{10}dx_{11}\ldots dx_{1n}\, dx_{20}dx_{21}\ldots dx_{2n} \exp[i/\hbar \int_t^{t+T} L(x_1, x_2, x'_1, x'_2, v, t_a)\, dt_a] \tag{7.14}$$

where A is a constant, and all integrals over the dx's ranges from $-\infty$ to ∞. Note that we use two sets of coordinates and momenta.

Following a similar path to subsection 7.1.1 we first we determine the fourier transform on the path integral for an infinitesimal time interval ε:

$$G_\varepsilon = \iint dx_1dx_2 \exp[-ip_1x_1 - ip_2x_2] \exp[-imx_1x_2/\varepsilon] \\ \sim \exp[-i\varepsilon p_1p_2/m] \tag{7.15}$$

Upon combining the intervals to a total time T we obtain the fourier transform of the total path integral

$$K(p, T) \sim \exp[-iTp_1p_2/m] \tag{7.16}$$

which yields the spatial path integral

$$K(x_1 - y_1, x_2 - y_2, T) \sim \int d\, p_1dp_2 \exp[ip_1(x_1 - y_1) + ip_2(x_2 - y_2)] \exp[-iTp_1p_2/m] \\ \sim \exp[-im(x_1 - y_1)(x_2 - y_2)/T] \tag{7.17}$$

which we normalize to unity

$$\int dy_1dy_2\, K(x_1 - y_1, x_2 - y_2, T) = 1 \tag{7.18}$$

with the result

$$K(x_1 - y_1, x_2 - y_2, T) = (m/T)\exp[-im(x_1 - y_1)(x_2 - y_2)/T] \tag{7.19}$$

If we now use the path integral on a fre particle wave function we see that it displaces the wave function by the time T:

$$\Psi_0(x_1, x_2, t) = \exp[-ip_1x_1 - ip_2x_2 - iEt] \qquad \text{where } E = p_1p_2/m \tag{7.20}$$

$$\Psi(y_1, y_2, t + T) = \iint dx_1 dx_2\, K(x_1 - y_1, x_2 - y_2, T)\Psi_0(x_1, x_2, t) \tag{7.21}$$
$$= \exp[-ip_1 y_1 - ip_2 y_2 - iE(t + T)]$$
$$= \Psi_0(y_1, y_2, t + T) \tag{7.22}$$

7.1.3 Introducing the Rotation Between the Quantum and Classical Cases of the Path Integral

We begin by expressing the coordinates in terms of new 'rotated' coordinates:

$$\begin{aligned} x_1 &= u_1 \cos\theta + u_2 \sin\theta \\ x_2 &= -u_1 \sin\theta + u_2 \cos\theta \\ p_1 &= -p_{u1} \sin\theta + p_{u2} \cos\theta \\ p_2 &= p_{u1} \cos\theta + p_{u2} \sin\theta \end{aligned} \tag{7.23}$$

Then the quantities of interest are now expressed as

$$\Psi_0(x_1, x_2, t) = \exp[-ip_1x_1 - ip_2x_2 - iEt] \qquad \text{where} \quad E = p_1p_2/m \tag{7.24}$$

$$= \Psi_0(u_1, u_2, t) = \exp\{-i[(p_{u2}u_2 - p_{u1}u_1)\sin(2\theta) + (p_{u2}u_1 + p_{u1}u_2)\cos(2\theta)] - iEt\} \tag{7.25}$$

where the energy E now having the form

$$E = (-p_{u1}\sin\theta + p_{u2}\cos\theta)(p_{u1}\cos\theta + p_{u2}\sin\theta)/m$$
$$= [(p_{u2}^2 - p_{u1}^2)\sin(2\theta)]/(2m) + [p_{u2}p_{u1}\cos(2\theta)]/m \tag{7.26}$$

We will now examine the two special cases: $\theta = 0$ corresponding to classical mechanics and $\theta = \pi/4$ corresponding to quantum mechanics.

$\underline{\theta = 0}$

The wave equation in this case is

$$\Psi_0(u_1, u_2, t) = \exp\{-i(p_{u2}u_1 + p_{u1}u_2) - iEt\} \tag{7.27}$$

with

$$E = p_{u2}p_{u1}/m$$

The wave function has a 'classical' form as we showed in eq. 47 in appendix C.

$\underline{\theta = \pi/4}$

The wave equation in this case is

$$\Psi_0(u_1, u_2, t) = \exp\{-i(p_{u2}u_2 - p_{u1}u_1) - iEt\} \tag{7.28}$$

with

$$E = (p_{u2}^2 - p_{u1}^2)/(2m)$$

This wave function has a quantum form with a positive energy part and a negative energy part:

7.1.4 Free Path Integral with Rotation Between Classical and Quantum Mechanics

The incremental path integral factor, expressed in terms of u_1 and u_2, and then fourier transformed is:

$$G_\varepsilon (p_{u1}, p_{u2}) = \iint du_1 du_2 \exp\{-i[(p_{u2}u_2 - p_{u1}u_1)\sin(2\theta) + (p_{u2}u_1 + p_{u1}u_2)\cos(2\theta)]\} \cdot \cdot \exp\{-im[(u_2^2 - u_1^2)\sin(2\theta)/2 + u_2u_1\cos(2\theta)]\}/\varepsilon\}$$

$$\sim \exp\{-i\varepsilon[(p_{u2}^2 - p_{u1}^2)\sin(2\theta)/2 + p_{u1}p_{u2}\cos(2\theta)]/m\} \quad (7.29)$$

yielding the cumulative product for the time interval T

$$K(p_{u1}, p_{u2}, T) \sim \exp[-iT[(p_{u2}^2 - p_{u1}^2)\sin(2\theta)/2 + p_{u1}p_{u2}\cos(2\theta)]/m] \quad (7.30)$$

Upon fourier transforming to coordinate space we find

$$K(u_1 - v_1, u_2 - v_2, T) \sim \int dp_{u1}dp_{u2} \exp\{i[(p_{u2}w_2 - p_{u1}w_1)\sin(2\theta) + (p_{u2}w_1 + p_{u1}w_2)\cos(2\theta)]\}K(p_{u1}, p_{u2}, T)$$

$$\sim \exp[im[(w_2^2 - w_1^2)\sin(2\theta)/2 + w_2w_1\cos(2\theta)]/T] \quad (7.31)$$

where

$$w_i = u_i - v_i$$

The special cases of interest are:

<u>$\theta = 0$</u>

$$K(u_1 - v_1, u_2 - v_2, T) \sim \exp[imw_2w_1]/T] \quad (7.32)$$

This gives the *Classical* path integral without use of any limiting or approximation procedure such as one often finds in the literature.

<u>$\theta = \pi/4$</u>

This case yields the familiar free particle path integral – but with a part for a positive energy particle and a part for a negative energy particle. The negative energy part can be removed easily yielding the conventional free particle quantum path integral.

$$K(u_1 - v_1, u_2 - v_2, T) \sim \exp[im(w_2^2 - w_1^2)/(2T)] \quad (7.33)$$

7.2 General PseudoQuantum Formulation

The propagator K(x – y, t) for the conventional Feynman path integral formulation has the form:

$$K(x-y, T) = A \lim_{n\to\infty} \iiiint_{-\infty}^{+\infty} \ldots \int dx_0 dx_1 \ldots dx_n \exp[i/\hbar \int_t^{t+T} L(x, v, t_a)\, dt_a] \quad (7.34)$$

where A is a constant, and the integral over the dx's ranges from –∞ to ∞.

The PseudoQuantum form of the path integral formalism is based on the PseudoQuantum lagrangian

$$L(x_1, v_1, x_2, v_2) = \partial^2 H(x_1, p_1)/\partial p_1^{\,2}\, p_1 p_2 + \partial^2 H(x_1, p_1)/\partial x_1 \partial p_1\, p_1 x_2 - \partial H(x_1, p_1)/\partial x_1\, x_2 \quad (7.35)$$

$$K(x_1 - y_1, x_2 - y_2, T) = A \lim_{n\to\infty} \iiiint_{-\infty}^{+\infty} \ldots \int dx_{10} dx_{11} \ldots dx_{1n}\, dx_{20} dx_{21} \ldots dx_{2n} \exp[i/\hbar \int_t^{t+T} L(x_1, v_1, x_2, v_2)\, dt_a] \quad (7.36)$$

From subsection 7,1.2, where

$$G_\varepsilon = \iint dx_1 dx_2 \exp[-ip_1x_1 - ip_2x_2] \exp[-imx_1x_2/\varepsilon] \sim \exp[-i\varepsilon p_1p_2/m] \quad (7.37)$$

we can perform the x_2 integration in G_ε using

$$x'_2 \cong x_2(t+\varepsilon) - x_{2(}(t)/\varepsilon \quad (7.38)$$

if

$$p_2 = mx'_2$$

Then

$$G_\varepsilon = \iint dx_1 dx_2 \exp[-ip_1x_1 - ip_2x_2] \exp\{-i\varepsilon[\partial^2 H(x_1, p_1)/\partial p_1^{\,2}\, p_1 m x_2/\varepsilon + \partial^2 H(x_1, p_1)/\partial x_1 \partial p_1\, p_1 x_2 - \partial H(x_1, p_1)/\partial x_1\, x_2]]$$

$$= \int dx_1 \exp[-ip_1x_1]\, \delta(p_2 - (m\partial^2 H(x_1, p_1)/\partial p_1^{\,2}\, p_1 + \varepsilon\partial^2 H(x_1, p_1)/\partial x_1\partial p_1\, p_1 - \varepsilon\partial H(x_1, p_1)/\partial x_1))$$

$$= \int dx_1 \exp[-ip_1x_1]\, \delta(p_2 - (m\partial^2 H(x_1, p_1)/\partial p_1^{\,2}\, p_1 + \varepsilon\partial^2 H(x_1, p_1)/\partial x_1\partial p_1\, p_1 - \varepsilon\partial H(x_1, p_1)/\partial x_1)) \quad (7.39)$$

7.2.1 PseudoQuantum Wave Functions and Schrödinger equation

In the presence of a potential, the path integral formulation leads us to transform it into a Schrödinger equation. The incremental time displacement form of the wave equation is

$$\Psi(y_{1k+1}, y_{2k+1}, t+\varepsilon) = \iint dx_{1k} dx_{2k} \exp\{(i\varepsilon/\hbar)[m(x_{1k+1} - x_{1k})(x_{2k+1} - y_{2k})/\varepsilon^2 - - x_{2k+1}(\partial H(x_1, p_1)/\partial x_1)|_{x_1 = x_{1k+1}}]\}\Psi_0(x_{1k}, x_{2k}, t) \quad (7.40)$$

$$= \iint dx_{1k}dx_{2k} \exp\{(i\varepsilon/\hbar)[m(x_{1k+1} - x_{1k})(x_{2k+1} - y_{2k})/\varepsilon^2 - - x_{2k+1}(\partial V(x)/\partial x)|_{x=x_{1k+1}}]\}\Psi_0(x_{1k}, x_{2k}, t)$$

for a potential V(x). In the case of the harmonic oscillator the PseudoQuantum potential term is

$$x_{2k+1}\partial V(x)/\partial x)|_{x=x_{1k+1}} = m\omega^2 x_{1k+1}x_{2k+1} \tag{7.41}$$

The PseudoQuantum Schrödinger equation that results is

$$i\hbar\partial\Psi(x_1, x_2, t)/\partial t = (-\hbar^2/m)\partial^2\Psi(x_1, x_2, t)/\partial x_1\partial x_2 + V(x_1, x_2)\Psi(x_1, x_2, t) \tag{7.42}$$

7.2.2 Decomposition of PseudoQuantum Schrödinger Equation into Quantum and Classical Parts

The Schrödinger equation can be decomposed into a quantum and a classical part. Starting from eq. 7.42:

$$i\hbar\partial\Psi(x_1, x_2, t)/\partial t = (-\hbar^2/m)\partial^2\Psi(x_1, x_2, t)/\partial x_1\partial x_2 + V(x_1, x_2)\Psi(x_1, x_2, t) \tag{7.43}$$

we find

$$i\hbar\partial\Psi(u_1, u_2, t, \theta)/\partial t = (-\hbar^2/m)[\sin(2\theta)\,(\partial^2/\partial u_2^2 - \partial^2/\partial u_1^2)\,/2 + \cos(2\theta)\partial^2/\partial u_1\partial u_2]\Psi(u_1, u_2, t, \theta) + + V(u_1 \cos\theta + u_2 \sin\theta, -u_1 \sin\theta + u_2 \cos\theta)\Psi(u_1, u_2, t, \theta) \tag{7.44}$$

using the relation to u_1 and u_2:

$$x_1 = u_1 \cos\theta + u_2 \sin\theta$$
$$x_2 = -u_1 \sin\theta + u_2 \cos\theta$$

Then

$$\Psi(u_1, u_2, t, \theta) = \exp\{-i[(p_{u2}u_2 - p_{u1}u_1)\sin(2\theta) + (p_{u2}u_1 + p_{u1}u_2)\cos(2\theta)] - iEt\} \tag{7.45}$$

and the energy is

$$E = [(p_{u2}^2 - p_{u1}^2)\sin(2\theta)]/(2m) + [p_{u2}p_{u1}\cos(2\theta)]/m \tag{7.46}$$

Again there are two cases of interest:

<u>The Quantum part $\theta = \pi/4$</u>

$$i\hbar\partial\Psi(u_1, u_2, t, \theta = \pi/4)/\partial t = (-\hbar^2/m)[(\partial^2/\partial u_2^2 - \partial^2/\partial u_1^2)\,/2]\Psi(u_1, u_2, t, \theta = \pi/4) + + V((u_1 + u_2)/\sqrt{2}, (-u_1 + u_2)/\sqrt{2})\Psi(u_1, u_2, t, \theta = \pi/4) \tag{7.47}$$

In the quantum free particle case

$$\Psi(u_1, u_2, t, \theta = \pi/4) = \exp\{-i(p_{u2}u_2 - p_{u1}u_1) - iEt\} \tag{7.48}$$

where

$$E = (p_{u2}^2 - p_{u1}^2)/(2m)$$

<u>The Classical part $\theta = 0$</u>

$$i\hbar\partial\Psi(u_1, u_2, t, \theta = 0)/\partial t = (-\hbar^2/m)\partial^2/\partial u_1\partial u_2\,\Psi(u_1, u_2, t, \theta = 0) + V(u_1, u_2)\Psi(u_1, u_2, t, \theta = 0)$$

In the classical free particle case

$$\Psi(u_1, u_2, t, \theta = 0) = \exp\{-i(p_{u2}u_1 + p_{u1}u_2) - iEt\} \tag{7.49}$$

where

$$E = p_{u2}p_{u1}/m$$

7.3 Classical Part of Free Particle PseudoQuantum Feyman Path Propagator

The classical part of the free particle PseudoQuantum path integral is described as follows. First the time incremental factor is

$$G_\varepsilon(p_{u1}, p_{u2}) = \iint du_1du_2 \exp\{-i[(p_{u2}u_2 - p_{u1}u_1)\sin(2\theta) + (p_{u2}u_1 + p_{u1}u_2)\cos(2\theta)]\} \cdot$$
$$\cdot \exp\{-im[(u_2^2 - u_1^2)\sin(2\theta)/2 + u_2u_1\cos(2\theta)]\}/\varepsilon\}$$

$$\sim \exp\{-i\varepsilon[(p_{u2}^2 - p_{u1}^2)\sin(2\theta)/2 + p_{u1}p_{u2}\cos(2\theta)]/m\} \tag{7.50}$$

The product of the incremental terms is

$$K(p_{u1}, p_{u2}, T) \sim \exp[-iT[(p_{u2}^2 - p_{u1}^2)\sin(2\theta)/2 + p_{u1}p_{u2}\cos(2\theta)]/m] \tag{7.51}$$

which yields the *classical* path integral

$$K(u_1 - v_1, u_2 - v_2, T) \sim \int dp_{u1}dp_{u2} \exp\{i[(p_{u2}w_2 - p_{u1}w_1)\sin(2\theta) + (p_{u2}w_1 + p_{u1}w_2)\cos(2\theta)]\}K(p_{u1}, p_{u2}, T)$$
$$\sim \exp[im[(w_2^2 - w_1^2)\sin(2\theta)/2 + w_2w_1\cos(2\theta)]/T] \tag{7.52}$$

where

$$w_i = u_i - v_i$$

For $\theta = 0$ the classical path integral steps are:

$$G_\varepsilon (p_{u1}, p_{u2}) \sim \exp\{-i\varepsilon p_{u1}p_{u2}/m\} \tag{7.53}$$

$$K(p_{u1}, p_{u2}, T) \sim \exp[-iTp_{u1}p_{u2}/m] \tag{7.54}$$

$$K(u_1 - v_1, u_2 - v_2, T) \sim \exp[im(u_1 - v_1)(u_2 - v_2)/T] \tag{7.55}$$

This gives us a classical path integral formulation that avoids approximation techniques which have hitherto been used. The above development can be rewritten in terms of the x and p variables.

7.4 Fokker-Planck Equation

The Feynman path integral formulation and equations can be transformed into similar forms by letting $i\hbar$ be changed to a positive constant. The Fokker-Planck equation gives the probability density of a particle velocity's time evolution under the impact of forces. This equation is also known as the Smoluchowski equation—named after its originator.

In its formulation a variable x_2 is introduced as the 'response' variable to the 'primary' variable x_1.The Fokker-Planck equation for the probability density is

$$p(x_1', t + \varepsilon) = (1/2\pi i) \int_{-\infty}^{\infty} dx_1 \int_{-i\infty}^{i\infty} dx_2 \exp\{\varepsilon[-x_2(x_1' - x_1)/\varepsilon + x_2D_1(x_1, t) + x_2^2D_2 (x_1, t)]\}p(x_1, t) \tag{7.56}$$

The origin of x_2 in our formalism, and in the Fokker-Planck equation, are very different. The Fokker-Planck equation has the lagrangian:

$$L = \int dt\, [x_2D_1(x_1, t) + x_2^2D_2 (x_1, t) - x_2\, \partial x_1/\partial t]$$

A comparison of this formulation with the PseudoQuantum path integral formulation shows an apparently remarkable similarity. Thus one might view the Fokker-Planck equation as a precursor of the PseudoQuantum formulation. The papers appearing in the appendices of this book show that the origin of our formalism and the Fokker-Planck formalism are very different.

8. The Transition Between Classical and Quantum Chaos

Chaos has become an increasingly important field of activity. While classical chaos has been the more studied aspect of chaos there has been an increasingly larger interest in quantum chaos. In this chapter we wish to show that the PseudoQuantum formalism appears to be a useful means of relating classical chaos to quantum chaos for many systems. It makes it possible to trace the transition from a classical chaos situation to quantum chaos. It also offers the possibility to determine the quantum analogue of a classical chaos phenomenon.

Given a quantum system it is often difficult to determine whether it has a chaotic regime. Frequently extensive numerical analysis is needed for this determination.

In this chapter we show that the PseudoQuantum formalism enables us to determine a classical Hamiltonian from a quantum formalism where chaos is known to occur based on extensive numerical investigations.

A well-studied[34,35] Hamiltonian for a quantum theory, known to have chaotic regions, is

$$H = (p_x^2 + p_y^2/2 + x^2y^2 + \beta(x^4 + y^4)/4 \qquad (8.1)$$

Creating the equivalent PseudoQuantum hamiltonian we obtain

$$\hat{H} = p_{x1}p_{x2} + p_{y1}p_{y2} + y_1^2x_1x_2 + x_1^2y_1y_2 + \beta(x_1^3x_2 + y_1^3y_2) \qquad (8.2)$$

Introducing new variables we can develop a form of eq. 8.2 that allows us to trace the transition from a quantum theory to a classical theory, which also should have chaotic regimes.

$$x_1 = u_{x1} \cos\theta + u_{x2} \sin\theta \qquad (8.3)$$
$$x_2 = -u_{x1} \sin\theta + u_{x2} \cos\theta$$

$$p_{x1} = p_{ux1} \cos\theta + p_{ux2} \sin\theta$$
$$p_{x2} = -p_{ux1} \sin\theta + p_{ux2} \cos\theta$$

---

[34] The Hamiltonian model above has been studied by: Y. Y. Bai, G. Hose, K. Stefański, and H. S. Taylor, Phys. Rev. **A31**, 2821 (1985), R. L. Waterland, J.-M. Yuan, C. C. Martens, R. E. Gillilan, and W. P. Reinhardt, Phys. Rev. Lett. **61**, 2733 (1988, and other papers..

[35] Another much studied model—the 2-Dimensional stadium, Quantum billiard ball model has classical chaotic dynamics, and a quantum approximation that is chaotic: S. W. McDonald and A. N. Kaufman, Phys. Rev. **A37**, 3067 (1988); ________, Phys. Rev. Lett., **42**, 1189 (1979), and other papers.

$$y_1 = u_{y1} \cos\theta + u_{y2} \sin\theta$$
$$y_2 = -u_{y1} \sin\theta + u_{y2} \cos\theta$$

$$p_{y1} = p_{uy1} \cos\theta + p_{uy2} \sin\theta$$
$$p_{y2} = -p_{uy1} \sin\theta + p_{uy2} \cos\theta$$

Then we obtain the PseudoQuantum Hamiltonian

$$\hat{H}(\theta) = p_{x1}p_{x2} + p_{y1}p_{y2} + y_1^2x_1x_2 + x_1^2y_1y_2 + \beta(x_1^3x_2 + y_1^3y_2)$$

$$= (p_{ux2}^2 - p_{ux1}^2 + p_{uy2}^2 - p_{uy1}^2)\sin(2\theta)/2 + (p_{uy1}p_{uy2} + p_{ux1}p_{ux2})\cos(2\theta) + (u_{y1} \cos\theta + u_{y2} \sin\theta)^2[(u_{x2}^2 - u_{x1}^2)\sin(2\theta)/2 + u_{x1}u_{x2} \cos(2\theta)] + (u_{x1} \cos\theta + u_{x2} \sin\theta)^2[(u_{y2}^2 - u_{y1}^2)\sin(2\theta)/2 + u_{y1}u_{y2} \cos(2\theta)] + \beta\{(u_{x1} \cos\theta + u_{x2} \sin\theta)^2[(u_{x2}^2 - u_{x1}^2)\sin(2\theta)/2 + u_{x1}u_{x2} \cos(2\theta)] + (u_{y1} \cos\theta + u_{y2} \sin\theta)^2[(u_{y2}^2 - u_{y1}^2)\sin(2\theta)/2 + u_{y1}u_{y2} \cos(2\theta)]\} \quad 8.4)$$

The classical Hamiltonian that results from this analysis is

$$\hat{H}(0) = (p_{uy1}p_{uy2} + p_{ux1}p_{ux2}) + u_{y1}^2u_{x1}u_{x2} + u_{x1}^2u_{y1}u_{y2} + \beta\{u_{x1}^2u_{x1}u_{x2} + u_{y1}^2u_{y1}u_{y2}\} \quad (8.5)$$

The Quantum Hamiltonian that emerges is

$$\hat{H}(\pi/4) = (p_{ux2}^2 - p_{ux1}^2 + p_{uy2}^2 - p_{uy1}^2)/2 + u_{y2}^2(u_{x2}^2 - u_{x1}^2)/2 + u_{x2}^2(u_{y2}^2 - u_{y1}^2)/2 + \beta\{u_{x2}^2(u_{x2}^2 - u_{x1}^2)/2 + u_{y2}^2(u_{y2}^2 - u_{y1}^2)/2\} \quad (8.6)$$

While these Hamiltonians require numerical analysis to understand their chaotic features, they offer the possibilities of comparative studies of quantum and classical chaos.

The study of other models of a similar character would appear to be of importance in elucidating quantum chaos.

9. The Transition Between Classical & Quantum Entanglement Dynamics

Quantum Entanglement has become of great importance judging from its increasing number of papers. It offers the possibility of new forms of communication that might be of great value in interstellar communication should Mankind reach the stars.

In this chapter we will study a prototype example of quantum entanglement with a view towards investigating the transition from quantum entanglement through 'semi-classical' quantum entanglement to a 'classical' limit.

The example that we consider will use a combination of a positive energy single particle state entangled with a negative energy particle state to simulate the more commonly studied case of entangled spins. The particles, in a superposed state, are assumed to separate, and one particle will be measured thus determining the state of the other particle due to entanglement. We define the 'entangled' NOON-type state:

$$\Psi = (|n+ = 1, n- = 0> + | n+ = 0, n- = 1>)/\sqrt{2} \qquad (9.1)$$

where one pure state $|n+ = 1, n- = 0>$ is a one particle state of positive energy and the other state $| n+ = 0, n- = 1>$ has a one negative energy particle.

We define projection operators of the form:[36]

$$\rho(\theta) = ||n+(\theta),n-(\theta)><n+(\theta),n-(\theta)| \qquad (9.2)$$

using an angle θ as we have done previously to specify the quantum-classical content of the projection.

More generally we will define harmonic oscillator states:

$$| n+, n-> = b_1^{\dagger n_+} b_2^{\dagger n_-}|0,0> \qquad (9.3)$$

with a density operator

$$\rho(\theta) = \sum_{n+,n-} |n+, n->< n+, n-| = \sum_{n+,n-} b_1^\dagger(\theta)^{n_+} b_2^\dagger(\theta)^{n_-}|0,0><0,0|b_1(\theta)^{n_+} b_2(\theta)^{n_-} \qquad (9.4)$$

We will consider the particular projection:

[36] We will be using the harmonic oscillator formalism of chapter 4 although the conclusions will be more far reaching.

$$P(\theta) = |1(\theta),0\rangle\langle 1(\theta),0| \quad (9.5)$$

which we will apply to Ψ:

$$P(\theta)\Psi = |1(\theta), 0\rangle/\sqrt{2} = \sin(2\theta)b_1^{\dagger}(\theta)|0, 0\rangle/\sqrt{2} = \sin(2\theta)(Q_1\cos\theta - iP_2\sin\theta)|0, 0\rangle/\sqrt{2} \quad (9.6)$$

Using the familiar relations:

$$b_1 = Q_1\cos\theta + iP_2\sin\theta \quad (4.6)$$
$$b_2 = -Q_2\sin\theta + iP_1\cos\theta$$

$$b_1^{\dagger} = Q_1\cos\theta - iP_2\sin\theta \quad (4.7)$$
$$b_2^{\dagger} = -Q_2\sin\theta - iP_1\cos\theta$$

with commutation relations

$$[b_1, b_1^{\dagger}] = \sin(2\theta) \quad (4.8a)$$
$$[b_2, b_2^{\dagger}] = -\sin(2\theta)$$
$$[b_1, b_2^{\dagger}] = [b_2, b_1^{\dagger}] = 0$$
$$[b_1, b_2] = [b_1^{\dagger}, b_2^{\dagger}] = 0$$

We find

$$\Psi' = P(\theta)\Psi \quad (9.7)$$

has the following forms for $\theta = \pi/4$ and $\theta=0$:

$\theta = \pi/4$

$$\Psi' = (Q_1 - iP_2)|0, 0\rangle/2 \quad (9.8)$$

gives a quantum state.

$\theta = 0$

$$\Psi' = 0 \quad (9.9)$$

We note that we showed in chapter 4 that 'classical' states containing particles have infinite energy—This above result, $\Psi' = 0$, reflects the infinite energy of states containing one or more particles. The value of $\Psi' = 0$ leaves the positivity or negativity of the other particle undetermined since entanglement is a purely quantum phenomena.

Other values of θ

$$\Psi' = \sin(2\theta)(Q_1\cos\theta - iP_2\sin\theta)|0, 0\rangle/\sqrt{2} \quad (9.10)$$
$$= \sin(2\theta)b_1^{\dagger}(\theta)|0, 0\rangle/\sqrt{2} = \sin(2\theta)|1(\theta),0\rangle/\sqrt{2}$$

An intermediate result occurs but the other entangled separated particle state's energy is determined to be negative.

10. PseudoQuantum Transition Between Quantum and [37]Classical Dynamics

The preceding chapters have shown that the use of the PseudoQuantum framework, which contains a purely quantum sector (albeit with both positive and negative energy parts that are separable), a purely classical sector, and an intermediate sector that is partly quantum and partly classical, enables us to

1. Relate the corresponding quantum and classical dynamics of a physical phenomenon.

2. Study the transition from quantum to classical behavior without recourse to approximations or limits such as $\hbar \to 0$.

3. Determine the classical equivalent of a quantum dynamical system.

4. Determine the quantum equivalent of a classical dynamical system.

These advantages appear to be fairly general in nature—as evidenced by the harmonic oscillator case studied in chapter 4—since the harmonic oscillator plays such a prominent role in many physical situations.

We conclude that the PseudoQuantum formalism, which is not only relevant for quantum-classical mechanics dynamics, but is also of importance in Quantum Field Theories as shown in our earlier chapters and in our papers in the appendices. It has also recently been used in a new GraviStrong unified theory that relates quark confinement to deviations from Newtonian gravitation at galactic distance scales using a canonical PseudoQuantum formulation of a higher derivative Quantum Field Theory. (See Blaha (2016e).)

---

[37] Gutzwiller (1990) points out the use of harmonic oscillator wave functions in several studies of the quantum-classical connection.

Appendix A. The Local Definition of Asymptotic Particle States

This refereed paper is S. Blaha, Il Nuovo Cimento **49 A**, 35 (1979). Reprinted with the kind permission of Il Nuovo Cimento.

IL NUOVO CIMENTO VOL. 49 A, N. 1 1 Gennaio 1979

The Local Definition of Asymptotic Particle States (*).

S. BLAHA

Physics Department, Williams College - Williamstown, Ma. 01267

(ricevuto il 28 Luglio 1978)

Summary. — A generalization of quantum field theory is described which has a unique particle interpretation even in space-times where no global timelike co-ordinate exists. The formulation is described in detail for the case of scalar bosons and spin-one-half fermions in flat space-time. We show that it is possible to construct a model in our approach which is physically equivalent to any given model in the usual formulation. In addition, a new class of models can be constructed which are not possible in the usual formulation. This class includes quantum action-at-a-distance models which can be used to develop models with higher-derivative field equations which are unitary. Our formulation allows some latitude in the choice of boundary conditions, so that one can opt for a continuum of possible Green's functions ranging from Feynman propagators to principal-value propagators (half advanced-half retarded).

1. – Introduction.

Our experience in flat space-time has fostered the opinion that a given action leads to a unique quantum field theory upon implementation of the canonical quantization procedure. This is apparently not true in general. A given action corresponds to an infinity of physically inequivalent quantum field theories in nonstatic space-times where no timelike Killing vector exists [1,2]. The origin of this plurality of quantum theories can be seen in free field the-

(*) Supported in part by grants from the National Science Foundation, and Research Corporation.

(1) S. A. FULLING: *Phys. Rev. D*, **7**, 2850 (1973); C. SOMMERFIELD: *Ann. Phys.*, **84**, 285 (1974).

(2) B. DEWITT: *Phys. Rep.*, **19**, 295 (1975).

ories (cf. FULLING [1]). The usual quantization procedure is based on a definition of positive frequency which selects an acceptable complete orthonormal set of field equation solutions to use in field quantization. In nonstatic space-times no unique criterion exists for defining positive frequency. As a result, there is no restriction on the choice of complete orthonormal set of field equation solutions used to Fourier-expand fields. Having different choices leads to unitarily (and physically) inequivalent representations of the field algebra. The set of physical particle states in one quantization is generally not unitarily related to the set of physical states in another quantization [1].

The absence of a criterion to select the « correct » quantum-field theory in the usual formulation has led us to consider a generalization of quantum field theory. In this generalization we introduce extra degrees of freedom in such a way that quantizations based on differing definitions of positive frequency are unitarily equivalent. Thus for a given action there is one resulting quantum field theory up to unitary equivalence.

In particular the physical particle states of different quantizations are related by a unitary transformation. Since the particle number operator is invariant under this transformation, a N-particle state in one quantization is a superposition of N-particle states in any other quantization. This is made possible by a local definition of particle states in the Fourier-transformed space (momentum space in the case of flat space-time).

It is important to note that the plethora of inequivalent quantizations in the usual formulation is faced by *one* observer. It is not a question of quantizations in different co-ordinate systems corresponding to different observers. The differences in the quantizations of two relatively accelerating observers, for example, are physically real and, in fact, also exist within the framework of our formulation. Relatively accelerating observers will, in general, « see » different numbers of particles.

Sections **2** and **3** contain our formulation of a free-scalar-boson field theory and a free spin-one-half fermion field theory in flat space-time. Significant differences exist between our formulation and the usual formulation. However, models exist in our formulation which make predictions which are identical to those of conventional field theory models, *e.g.*, quantum electrodynamics. Models also exist in our formulation which are completely outside the framework of the usual formulation. For example, a choice of boundary conditions is possible in our formulation which allows for virtual particles to propagate via non-Feynman propagators. In general, our particle propagator has the form

$$G = \sin^2\theta\, G_F + \cos^2\theta\, C G_F^* C^{-1}\,, \tag{1}$$

where θ is an arbitrary angle, G_F the usual Feynman propagator with G_F^* its complex conjugate, and where C is the relevant charge conjugation matrix. G is a Feynman propagator if $\theta = \pi/2$.

If $\theta = \pi/4$ then G is a principal-value propagator (half advanced-half retarded). This type of Green's function has appeared in classical action-at-a-distance theories. Our formulation thus encompases quantum action-at-a-distance theories. The use of principal-value propagators allows a substantial enlargement of the class of unitary, renormalizable field theory models. For example, models with higher-derivative field equations cannot simultaneously satisfy the requirements of positive probabilities and unitarity, if Feynman propagators are used. But if principal-value propagators are used, both requirements can be consistently satisfied (3). This has allowed us to previously construct a unitary, higher-derivative non-Abelian model of the strong interactions with a manifest linear potential and quark confinement (4). Of course, the use of principal-value propagators leads to a different type of analytic structure for amplitudes. We take the view that analyticity is an experimental question rather than a fundamental requirement on field theory (5). It is amusing to note that confinement of color in this model serves to sharply dampen if not eliminate the potential nonanalyticity.

We will discuss our formulation of non-Abelian field theories in detail in a subsequent paper (6).

2. – Boson quantization.

In flat space-time a timelike co-ordinate exists and as a result the Hamiltonian occupies a privileged position in defining positive frequency. In a nonstatic space-time, with no global timelike co-ordinate, no corresponding operator exists and the definition of « positive frequency » appears to be arbitrary. Consider a free scalar-field theory in such a situation. The field equation has an infinite number of possible complete orthonormal sets of solutions which span the space of solutions. Consider two possible sets: $\{\chi_\alpha, \chi_\alpha^*\}$ and $\{\psi_\beta, \psi_\beta^*\}$, where the χ_α are positive frequency with respect to one definition of positive frequency, and ψ_β are positive frequency with respect to a different definition. Then mode expansions of the scalar field

$$\varphi(x) = \sum_\alpha [\chi_\alpha(x) A_\alpha + \chi_\alpha^*(x) A_\alpha^\dagger], \tag{2}$$

$$\varphi(x) = \sum_\beta [\psi_\beta(x) b_\beta + \psi_\beta^*(x) b_\beta^\dagger], \tag{3}$$

(3) S. BLAHA: *Phys. Rev. D*, **10**, 4268 (1974).
(4) S. BLAHA: *Phys. Rev. D*, **11**, 2921 (1975).
(5) R. E. CUTKOWSKY, P. V. LANDSHOFF, D. I. OLIVE and J. C. POLKINGHORNE: *Nucl. Phys.*, **12** B, 281 (1969); T. D. LEE: in *Quanta-Essays in Theoretical Physics Dedicated to Gregor Wentzel*, edited by P. G. O. FREUND, C. J. GOBEL and Y. NAMBU (Chicago, Ill., 1970); H. RECHENBERG and E. C. G. SUDARSHAM: *Nuovo Cimento*, **14** A, 299 (1973).
(6) S. BLAHA: *Nuovo Cimento*, **49** A, 58 (1978).

can be inverted to relate the Fourier coefficient operators

$$A_\alpha = \sum_\beta [C_{\alpha\beta} b_\beta + \tilde{C}_{\alpha\beta} b_\beta^\dagger] , \tag{4}$$

where $C_{\alpha\beta}$ and $\tilde{C}_{\alpha\beta}$ are c-number functions of α and β. Equation (4) shows that A_α is related to b_β and $b_\beta^\dagger$ through a local Bogoliubov transformation. As a result, the quantizations are, in general, not unitarily equivalent, have different vacua, and different particle interpretations [1,2]. The basis of this difficulty is the noncommutativity of Fourier coefficient operators and their Hermitian conjugates

$$[b_\beta, b_{\beta'}^\dagger] = \delta_{\beta\beta'} . \tag{5}$$

We shall propose a generalization of quantum field theory in which (in the free-field case) the Fourier coefficient operators and their Hermitian conjugates commute. In order to maintain the quantum character of the theory a supplementary field and the corresponding Fourier coefficient operators will be introduced. We shall confine our discussion to flat space-time in this section, and in sect. **3** which deals with free spin-one-half fermion quantization. In sect. **4** we discuss the particle interpretation of the formulation in nonstatic space-time.

Let us provisionally introduce the Lagrangian

$$\mathscr{L} = \partial_\mu \varphi_1 \partial^\mu \varphi_2 - \tfrac{1}{2} \partial_\mu \varphi_1 \partial^\mu \varphi_1 - m^2 \varphi_1 \varphi_2 + \tfrac{1}{2} m^2 \varphi_1^2 . \tag{6}$$

Following canonical procedures, we obtain the field equations

$$(\Box + m^2) \varphi_i = 0 , \tag{7}$$

for $i = 1, 2$ and the canonical momenta

$$\pi_1 = \dot{\varphi}_2 - \dot{\varphi}_1 \tag{8}$$

and

$$\pi_2 = \dot{\varphi}_1 , \tag{9}$$

which are taken to satisfy the canonical equal-time commutation relations

$$[\varphi_i(x), \pi_j(y)] = i\delta_{ij} \delta^3(\boldsymbol{x} - \boldsymbol{y}) , \tag{10}$$

$$[\varphi_i(x), \varphi_j(y)] = [\pi_i(x), \pi_j(y)] = 0 , \tag{11}$$

for $i, j = 1, 2$. Equations (9) and (10) imply

$$[\varphi_1(x), \dot{\varphi}_1(y)] = 0\,, \tag{12}$$

$$[\varphi_1(x), \dot{\varphi}_2(y)] = i\delta^3(\boldsymbol{x} - \boldsymbol{y})\,, \tag{13}$$

$$[\varphi_2(x), \dot{\varphi}_2(y)] = i\delta^3(\boldsymbol{x} - \boldsymbol{y})\,, \tag{14}$$

at equal times. The most general form for the mode expansion of the fields is

$$\varphi_1(x) = \int d^3k\,[(C_{11}A_{1k} + C_{12}A_{2k})f_k(x) + (\tilde{C}_{11}A^\dagger_{1k} + \tilde{C}_{12}A^\dagger_{2k})f^*_k(x)]\,, \tag{15}$$

$$\varphi_2(x) = \int d^3k\,[(C_{21}A_{1k} + C_{22}A_{2k})f_k(x) + (\tilde{C}_{21}A^\dagger_{1k} + \tilde{C}_{22}A^\dagger_{2k})f^*_k(x)]\,, \tag{16}$$

where $(2\pi)^{\frac{3}{2}}(2\omega_k)^{\frac{1}{2}}f_k(x) = \exp[-ik\cdot x]$ and where C_{ij} and $\tilde{C}_{ij}$ are a set of constants. In view of the afore-mentioned difficulties stemming from the non-commutativity of a Fourier-coefficient operator and its Hermitian conjugate, we are led to impose the commutation relations

$$[A_{ik}, A_{jk'}] = [A^\dagger_{ik}, A^\dagger_{ik'}] = 0 \tag{17}$$

and

$$[A_{ik}, A^\dagger_{jk'}] = (1 - \delta_{ij})\delta^3(\boldsymbol{k} - \boldsymbol{k}')\,, \tag{18}$$

for $i, j = 1, 2$. We define two vacua (which are in fact related) $|0\rangle_1$ and $|0\rangle_2$ by

$$A_{1k}|0\rangle_2 = A^\dagger_{1k}|0\rangle_2 = 0\,, \tag{19}$$

$$A_{2k}|0\rangle_1 = A^\dagger_{2k}|0\rangle_1 = 0 \tag{20}$$

with

$$A_{2k}|0\rangle_2 \neq 0\,, \quad A^\dagger_{2k}|0\rangle_2 \neq 0\,, \tag{21}$$

and

$$A_{1k}|0\rangle_1 \neq 0\,, \quad A^\dagger_{1k}|0\rangle_1 \neq 0\,, \tag{22}$$

for all k. These definitions are motivated by the need for vacua which would be invariant under Bogoliubov transformations—a necessary requirement if the difficulties of particle interpretation caused by relations of the form of eq. (4) are to be avoided. Let us define the local Bogoliubov transformation

$$A_{ik}(\lambda_1, \lambda_2) = B_{\lambda_1\lambda_2}A_{ik}B^{-1}_{\lambda_1\lambda_2} = \tag{23}$$

$$= \exp[i\lambda_1]\cosh\lambda_2 A_{ik} + \exp[-i\lambda_1]\sinh\lambda_2 A^\dagger_{ik}\,,$$

where λ_1 and λ_2 are functions of the momentum k. The operator B has the form

$$B_{\lambda_1\lambda_2} = \exp\left[2i\int d^3k\lambda_1(k)\Gamma_{3k}\right]\exp\left[2i\int d^3k\lambda_2(k)\Gamma_{2k}\right], \tag{24}$$

where

$$\Gamma_{3k} = (A^\dagger_{2k}A_{1k} + A_{2k}A^\dagger_{1k})/2\,, \tag{25}$$

$$\Gamma_{2k} = i(A^\dagger_{2k}A^\dagger_{1k} - A_{2k}A_{1k})/2\,. \tag{26}$$

If we also define

$$\Gamma_{1k} = -(A^\dagger_{2k}A^\dagger_{1k} + A_{2k}A_{1k})/2\,, \tag{27}$$

then these operators satisfy the commutation relations of a $SU_{1,1}$ algebra:

$$[\Gamma_{1k}, \Gamma_{2k'}] = -i\delta_{kk'}\Gamma_{3k}\,, \qquad [\Gamma_{2k}, \Gamma_{3k'}] = i\delta_{kk'}\Gamma_{1k}\,, \qquad [\Gamma_{3k}, \Gamma_{1k'}] = i\delta_{kk'}\Gamma_{2k}\,. \tag{28}$$

Thus the group of local Bogoliubov transformations is an infinite tensor product of $SU_{1,1}$ groups. It should be noted that $|0\rangle_2$ and $|0\rangle_1$ are invariant under this group. The equations of motion and equal-time commutation relations are also invariant under this group. These properties will enable us to show the uniqueness of the particle interpretation of our formulation in nonstatic space-time in sect. **4**. The Casimir operator for the k-th $SU_{1,1}$ algebra,

$$\Gamma^2_k = \Gamma^2_{3k} - \Gamma^2_{1k} - \Gamma^2_{2k}\,, \tag{29}$$

$$\Gamma^2_k = N_k(N_k + 2)\,, \tag{30}$$

allows us to identify the particle number operator (cf. sect. **4** below for its derivation)

$$N = \int d^3k N_k = \int d^3k [A^\dagger_{2k}A_{1k} - A_{2k}A^\dagger_{1k}]\,, \tag{31}$$

which is left invariant by the Bogoliubov transformations.

If we compare our formulation to the usual one at this stage, we see that the enlargement of the field algebra has allowed us to define a group of local Bogoliubov transformations which is unitary and leaves the vacuum invariant —two properties not possible in the usual approach.

We now define inner products in our formalism. The structure of the commutation relations eq. (17) and (18) together with the nature of the defined vacua suggest that kets can be taken to have the form

$$|\alpha\rangle = A^\dagger_{2k_1}A^\dagger_{2k_2}\dots A_{2q_1}A_{2q_2}\dots|0\rangle_2 \tag{32}$$

and that bras should have the form

$$\langle\alpha| = {}_1\langle 0|A_{1k_1}A_{1k_2}\dots A^\dagger_{1q_1}A^\dagger_{1q_2}\dots\,. \tag{33}$$

(We could have chosen to construct kets using $|0\rangle_1$ and bras using ${}_2\langle 0|$ with no change in consequences.) The form of the commutation relations (which are used to reduce the inner products to a multiple of ${}_1\langle 0|0\rangle_2$) imply that the dual of the ket space is not its Hermitian conjugate. In our case the algebra reduces inner products to ${}_1\langle 0|0\rangle_2$ which we define to be unity.

We can relate the dual of a ket to its Hermitian conjugate through the introduction of a Dirac metric operator ([8]). We define the operator, γ, by

$$\gamma^{-1}A_{1k}\gamma = A_{2k}\,, \qquad \gamma^{-1}A_{2k}\gamma = A_{1k}\,, \tag{34}$$

$$\gamma|0\rangle_1 = |0\rangle_2\,. \tag{35}$$

We find

$$\gamma = \exp\left[-\frac{i\pi}{2}\int \mathrm{d}^3k\,[A_{2k}^\dagger A_{2k} + A_{1k}^\dagger A_{1k} - A_{2k}^\dagger A_{1k} - A_{2k}A_{1k}^\dagger]\right], \tag{36}$$

which implies that γ satisfies the necessary conditions for a metric operator, $\gamma = \gamma^\dagger = \gamma^{-1}$. The norm of a state $|\alpha\rangle$ can thus be defined by

$$(|\alpha\rangle)^\dagger\gamma|\alpha\rangle \tag{37a}$$

and inner products will generally have the form

$$(|\beta\rangle)^\dagger\gamma|\alpha\rangle\,. \tag{37b}$$

The adjoint operator is defined by

$$A^* = \gamma^{-1}A^\dagger\gamma\,. \tag{38}$$

Physical observables must be self-adjoint, $A^* = A$. Self-adjoint operators play the same role as Hermitian operators do in the usual formulation. In particular the Hamiltonian must be self-adjoint, if we are to have conservation of norm. Because φ_2 satisfies a Jordan-Pauli commutation relation, we shall introduce interactions in our model using only $\varphi_2(x)$. As a result φ_2 must also be self-adjoint if the Hamiltonian is to be self-adjoint.

([7]) Earlier two field formalisms have been considered by G. Mie: *Ann. Phys. Lpz.*, **37**, 511 (1912); P. A. M. Dirac: *Comm. Dublin Inst. Advanced Studies*, **180** A, 1 (1942); W. Pauli: *Rev. Mod. Phys.*, **15**, 175 (1943); M. Froissart: *Suppl. Nuovo Cimento*, **14**, 197 (1959); T. D. Lee and G. C. Wick: *Phys. Rev. D*, **2**, 1033 (1970). Our motivation and formulation differ substantially from them. Ref. ([3,4]) above do describe models which can be directly incorporated within the framework of our formulation.

([8]) W. Pauli: *Rev. Mod. Phys.*, **15**, 175 (1943).

The energy-momentum tensor is defined by

$$T^{\mu\nu} = -g^{\mu\nu}\mathscr{L} + \frac{\delta\mathscr{L}}{\delta\partial_\mu\varphi_1}\partial^\nu\varphi_1 + \frac{\delta\mathscr{L}}{\delta\partial_\mu\varphi_2}\partial^\mu\varphi_2 \tag{39}$$

with the Hamiltonian given by the 0-0 component. It is easy to verify that the requirements of Poincaré invariance, and the Schwinger commutation relations for $T^{\mu\nu}$ are met.

The equal-time commutation relations, and the self-adjointness of H and φ_2 place six constraints on the constants C_{ij} and $\tilde{C}_{ij}$ in eqs. (15) and (16). After some algebra we find that we are able to express the field operators in the form

$$\varphi_1(x) = \int d^3k \left[\left(\frac{\cos(\theta_1-\theta_2)}{\sin\theta_1}A_{1k} + \frac{\sin(\theta_1-\theta_2)}{\sin\theta_1}A_{2k}\right)f_k(x) + \right.$$
$$\left. + \left(\frac{\cos(\theta_1-\theta_2)}{\cos\theta_1}A^\dagger_{1k} - \frac{\sin(\theta_1-\theta_2)}{\cos\theta_1}A^\dagger_{2k}\right)f^*_k(x)\right], \tag{40}$$

$$\varphi_2(x) = \int d^3k\left[(\cos\theta_2 A_{2k} + \sin\theta_2 A_{1k})f_k(x) + (\sin\theta_2 A^\dagger_{2k} - \cos\theta_2 A^\dagger_{1k})f^*_k(x)\right], \tag{41}$$

where θ_1 and θ_2 are arbitrary constants which fix the boundary conditions of the Green's functions. (They are *not* related to the Bogoliubov transformations defined above.) We also find

$$H = \int d^3k\,\omega_k(A^\dagger_{2k}A_{1k} + A_{2k}A^\dagger_{1k}) = 2\int d^3k\,\omega_k\Gamma_{3k} \tag{42}$$

in the free-field case independent of θ_1 and θ_2.

The theory is not invariant under Bogoliubov transformations due to their noncommutativity with H. This is consonant with the absence of any evidence in nature for such an invariance (and related constants of motion). The point of our formulation is to ensure that representations of the field algebra and dynamics, which are related to each other by Bogoliubov transformations, are unitarily equivalent. In the case of flat space-time the unitary equivalence is a moot point, since a unique generator of the dynamics, the Hamiltonian, is apparent. In nonstatic space-times, where no unique generator of the dynamical motion is determined, the unitary equivalence is necessary in order to have an unambiguous quantum field theory (given the action).

Different choices for the generator of the dynamics lead to representations which can be related by Bogoliubov transformations. These representations are unitarily equivalent in our formulation, but not equivalent in the usual formulation. We return to this issue in sect. **4**.

The role of θ_1 and θ_2 is evident in the Green's functions. As usual we define the Green's functions as the vacuum expectation values of the time-ordered product of the field operators:

$$iG_{ij}(x-y) = {}_1\langle 0|T(\varphi_i(x)\varphi_j(y))|0\rangle_2 \,. \tag{43}$$

Equation (41) implies

$$G_{22}(x-y) = \sin^2\theta_2 G_F(x-y) + \cos^2\theta_2 G_F^*(x-y)\,, \tag{44}$$

where $G_F(x-y)$ is the usual Feynman propagator. G_{12} and G_{11} also depend on θ_1 and θ_2, but their precise expressions will not be of use in our presentation.

We shall now show that a model exists within our formulation which is physically equivalent to any conventional scalar quantum field theory with interaction $\mathscr{L}_I(\varphi)$. Our model Lagrangian is given by eq. (6) plus the interaction Lagrangian $\mathscr{L}_I(\varphi_2)$, where $\mathscr{L}_I(\varphi_2)$ is the same function of φ_2 as $\mathscr{L}_I(\varphi)$ is of φ. In order to have Feynman propagators, it is necessary to choose the boundary condition $\theta_2 = \pi/2$. We shall demonstrate that an asymptotic state exists in our formulation which corresponds to any asymptotic state of the usual formulation, and then show that S-matrix elements between corresponding states in the two models are equal in any order of perturbation theory.

The construction of asymptotic fields and states in our model is based on the renormalized quadratic part of the Lagrangian (eq. (6)). Therefore the previous development of this section can be used if appropriate subscripts « in » or « out » are appended to the operators. In particular, since $\theta_2 = \pi/2$, we have

$$\varphi_{2\text{in}}(x) = \int d^3k\,[f_k(x)A_{1k\text{in}} + f_k^*(x)A_{2k\text{in}}^\dagger] \tag{45a}$$

by eq. (41). We shall express the in-field operator of the usual formulation by

$$\varphi_{\text{in}}(x) = \int d^3k\,[f_k(x)A_{k\text{in}} + f_k^*(x)A_{k\text{in}}^\dagger]\,. \tag{45b}$$

Note that the form of $\varphi_{2\text{in}}$ and φ_{in} is identical except for the subscripts « 1 » and « 2 » on the operators. In addition, the commutation relations of the field operators and the Fourier-coefficient operators are also identical except for numerical subscripts. Furthermore, the application of the field operators to the vacua is also identical in effect (except for subscripts)

$$\begin{cases} \varphi_{\text{in}}(x)|0\rangle = \varphi_{\text{in}}^{(-)}(x)|0\rangle\,, & \varphi_{2\text{in}}(x)|0\rangle_2 = \varphi_{2\text{in}}^{(-)}(x)|0\rangle_2\,, \\ \langle 0|\varphi_{\text{in}}(x) = \langle 0|\varphi_{\text{in}}^{(+)}(x)\,, & {}_1\langle 0|\varphi_{2\text{in}}(x) = {}_1\langle 0|\varphi_{2\text{in}}^{(+)}(x)\,, \end{cases} \tag{45c}$$

where the superscript « + » labels positive-frequency parts of the field operator and « − » labels negative-frequency parts. This close parallel in properties between our model and the model of the usual formulation implies the identity

$$\langle 0|\mathscr{P}(\varphi_{\text{in}})|0\rangle = {}_1\langle 0|\mathscr{P}(\varphi_{2\text{in}})|0\rangle_2\,, \tag{45d}$$

where $\mathscr{P}(\varphi_{\text{in}})$ is any polynomial in the field φ_{in}. Later we shall use this identity to demonstrate the equality of the S-matrices in our model and the given model of the usual formulation. (Note that a straightforward application of eq. (45*d*) implies that the propagator G_{22} in our formulation equals the time-ordered propagator of the usual formulation.)

We now state the rule associating asymptotic particle states in our formulation with those of the usual formulation: given an in or out ket of the usual formulation, the corresponding ket in our formulation is obtained by appending the subscript «2» to every Fourier-coefficient operator (and to the vacuum) (*e.g.* $A^{\dagger}_{k\text{in}}|0\rangle \Leftrightarrow A^{\dagger}_{2k\text{in}}|0\rangle_2$). Given an in or out bra of the usual formulation, the corresponding bra in our formulation is obtained by appending «1» to each Fourier-coefficient operator (and to the vacuum) (*e.g.* $\langle 0|A_{k\text{in}} \Leftrightarrow {}_1\langle 0|A_{1k\text{in}}$). It is easily seen that energy-momentum eigenstates in the usual formulation correspond to energy-momentum eigenstates in our formulation. Thus we have identified the set of physical states in our model and find a detailed correspondence to those of the usual formulation.

The development of the perturbation theory of our model is completely analogous to the usual development. The S-matrix relates in and out fields: $\varphi_{2\text{in}}(x) = S\varphi_{2\text{out}}(x)S^{-1}$. LSZ reduction formulae are derived in the same manner as in the usual formulation. We find the reduction formula for a particle from an in-state and from an out-state to be, respectively,

$$\text{(46)}\quad \left\{\begin{aligned} &\langle\beta\,\text{out}|\alpha p\,\text{in}\rangle = \\ &\qquad = \langle\beta - p\,\text{out}|\alpha\,\text{in}\rangle + \frac{i}{\sqrt{Z}}\int \mathrm{d}^4x\, f_p(x)(\overrightarrow{\Box + m^2})\langle\beta\,\text{out}|\varphi_2(x)|\alpha\,\text{in}\rangle\,, \\ &\langle\beta k\,\text{out}|\alpha\,\text{in}\rangle = \\ &\qquad = \langle\beta\,\text{out}|\alpha - k\,\text{in}\rangle + \frac{i}{\sqrt{Z}}\int \mathrm{d}^4x\, f_k^*(x)(\overrightarrow{\Box + m^2})\langle\beta\,\text{out}|\varphi_2(x)|\alpha\,\text{in}\rangle\,, \end{aligned}\right.$$

where $\varphi_2(x)$ is the interacting field and where we use the notation of ref. (9). The reduction of several particles leads to expressions which are identical to corresponding expressions of the usual model if the subscript «2» is appended to $\varphi(x)$.

Just as in the conventional model, we can formally develop a perturbation theory based on the U-matrix. The U-matrix relates the interacting and asymptotic field operator

$$\text{(47a)}\qquad \varphi_2(\boldsymbol{x}, t) = U^{-1}(t)\varphi_{2\text{in}}(\boldsymbol{x}, t)U(t)$$

(9) We follow the conventions and notation of J. D. BJORKEN and S. D. DRELL: *Relativistic Quantum Fields* (New York, N. Y., 1965).

and is easily shown to satisfy the differential equation

$$i\frac{\partial U}{\partial t} = -\left[\int d^3x\, \mathscr{L}_1(\varphi_{2\text{in}})\right]U\,. \tag{47b}$$

Defining $U(t,t') = U(t)U^{-1}(t')$ and solving eq. (47b) gives

$$U(t,t') = T\exp\left[i\int_{t'}^{t} d^4x\, \mathscr{L}_1(\varphi_{2\text{in}})\right]. \tag{48}$$

The LSZ procedure defined above reduces the calculation of S-matrix elements to the evaluation of time-ordered products of the interacting fields, ${}_1\langle 0|T(\varphi_2(x_1)\varphi_2(x_2)\dots\varphi_2(x_N))|0\rangle_2$. The U-matrix can then be used to reduce this quantity to the ratio of matrix elements involving only in-fields:

$$\frac{{}_1\langle 0|T(\varphi_{2\text{in}}(x_1)\dots\varphi_{2\text{in}}(x_N)\exp[i\int d^4x\, \mathscr{L}_1(\varphi_{2\text{in}})])|0\rangle_2}{{}_1\langle 0|T(\exp[i\int d^4x\, \mathscr{L}_1(\varphi_{2\text{in}})])|0\rangle_2}\,. \tag{49}$$

Expanding to any order in the interaction in eq. (49) gives matrix elements of polynomials in $\varphi_{2\text{in}}$ which are equal—term by term—to corresponding matrix elements of the perturbation theory of the model of the conventional formulation by eq. (45d). Thus S-matrix elements between corresponding states are identically equal in the conventional model and our corresponding model.

It should be noted that only a subset of the possible asymptotic states in our model are identified as physical particle states which correspond to states in the usual model. The operator $A_{2k\text{in}}$ can also be used to create in-kets (and $A^\dagger_{1k\text{out}}$ to create out bras), but the S-matrix elements between physical kets and any ket (or bra) in which these operators appear is zero. (This follows from the fact that $[\mathscr{L}_1(\varphi_{2\text{in}}), A_{2k\text{in}}] = [\mathscr{L}_1(\varphi_{2\text{in}}), A^\dagger_{1k\text{in}}] = 0$ and ${}_1\langle 0|A_{2k\text{in}} = 0 = A^\dagger_{1k\text{in}}|0\rangle_2$.) Thus the S-matrix is block diagonal in our model. The part of it corresponding to the physical state sector is identical to the S-matrix of the given model of the conventional formulation.

The expression for the vacuum expectation value from which S-matrix elements may be calculated, eq. (49), can be used to show the unitary equivalence of representations which are related by a Bogoliubov transformation. Suppose that we had not used the representation of eq. (45a), but instead the Bogoliubov-transformed representation

$$\varphi^{\text{B}}_{2\text{in}}(x) = \int d^3k\,[f_k(x)(A_{1k\text{in}}\cosh\lambda + A^\dagger_{1k\text{in}}\sinh\lambda) + f^*_k(x)(A^\dagger_{2k\text{in}}\cosh\lambda + A_{2k\text{in}}\sinh\lambda)] = B_{0\lambda}\varphi_{2\text{in}}(x)B^{-1}_{0\lambda}\,. \tag{50}$$

The canonical nature of the transformation guarantees that the canonical commutation relations will be maintained. If we follow the development of the perturbation theory given by eqs. (45)-(49) with φ^{B}_{2in} replacing φ_{2in} and $\varphi^{B}_{2} = U^{-1}(t)\varphi^{B}_{2in}U(t)$ replacing φ_2, then we find that S-matrix elements are calculated from vacuum expectation values involving only φ^{B}_{2in} fields:

$$(51)\qquad \frac{{}_1\langle 0|T(\varphi^{B}_{2in}(x_1)\ldots\varphi^{B}_{2in}(x_N)\exp[i\int d^4x\,\mathscr{L}_1(\varphi^{B}_{2in})])|0\rangle_2}{{}_1\langle 0|T(\exp[i\int d^4x\,\mathscr{L}_1(\varphi^{B}_{2in})])|0\rangle_2}\,.$$

Since $B^{-1}_{0\lambda}|0\rangle_2 = |0\rangle_2$ and ${}_1\langle 0|B_{0\lambda} = {}_1\langle 0|$ we find that eq. (51) is equal to eq. (49). Thus the unitary equivalence of representations of the quantum field theory differing by a Bogoliubov transformation is demonstrated. (One can formally define Bogoliubov transformations for interacting fields $B^{int} = UBU^{-1}$, but B^{int} is not unitary due to the well-known difficulties of the U-matrix in the conventional formulation which are also present in our formulation. We circumvent this problem by working with the definition of the S-matrix in terms of vacuum expectation values of asymptotic in-fields, where the unitary equivalence under Bogoliubov transformation can be unambiguously shown to hold.)

A comparison of our formulation with the usual formulation shows a certain similarity of form at the Lagrangian level if our Lagrangian is put in the form

$$(52)\qquad \mathscr{L} = -\frac{1}{2}\partial_\mu(\varphi_1-\varphi_2)\partial^\mu(\varphi_1-\varphi_2) + \partial_\mu\varphi_2\partial^\mu\varphi_2 + \\ + \mathscr{L}_1(\varphi_2) + \frac{m^2}{2}(\varphi_1-\varphi_2)^2 - \frac{m^2}{2}\varphi_2^2\,.$$

In the usual approach $\varphi_3 = \varphi_1 - \varphi_2$ is an ignorable field and it would not have been surprising that we found equal S-matrix elements above. However, our formulation differs from the usual formulation in two respects—first, the field operators are both expanded in type «1» and «2» Fourier coefficient operators and, more importantly, the vacuum is defined in a way which correlates the φ_3 and φ_2 sectors. In the $\theta_2 = \pi/2$ case the first difference can be eliminated by a relabeling of Fourier-coefficient operators. However, for other values of θ_2 both differences are present and lead to a very different theory from the usual formulation. While it is clear that the correlation between the φ_3 and φ_2 sectors can be implemented in free field theory, one might ask if this remains true in the interacting case. Certainly the correlation can be implemented in the asymptotic fields and states, since that is free field theory. One can also *formally* implement the correlation for interacting fields through eq. (47*a*). But the implementation of the correlation in the interacting case is actually based on the reduction of the S-matrix element to the vacuum expectation value of products of asymptotic fields. Since the correlation can be maintained for the asymptotic fields and states, the physical quantities of the models, S-matrix elements, embody the correlation. (The value of the correlation we introduce is twofold: first, it is necessary in order to obtain the unitary equivalence of Bogoliubov rotated representations, and secondly, it widens the range of allowed flat-space-time quantum field theories to include those with principal-value propagators ([6]).)

We conclude this section with a brief discussion of our formulation of the charged-scalar-particle case. In the usual approach, the free-charged-scalar-particle Lagrangian may be expressed in terms of complex fields, $\varphi(x)$ and $\varphi^*(x)$ or in terms of two real fields $\varphi_a(x)$ and $\varphi_b(x)$ with

$$\varphi(x) = [\varphi_a(x) + i\varphi_b(x)]/\sqrt{2}\,. \tag{53}$$

If we follow the same procedure as above for the real fields, double their number and use the Lagrangian form of eq. (6), we are led to the complex field expression of the Lagrangian:

$$\mathscr{L} = \partial_\mu\tilde{\varphi}_2\partial^\mu\varphi_1 + \partial_\mu\varphi_2\partial^\mu\tilde{\varphi}_1 - \partial_\mu\tilde{\varphi}_1\partial^\mu\varphi_1 - m_2\tilde{\varphi}_2\varphi_1 - m^2\varphi_2\tilde{\varphi}_1 + m^2\tilde{\varphi}_1\varphi_1\,, \tag{54}$$

where

$$\varphi_i(x) = [\varphi_{ia}(x) + i\varphi_{ib}(x)]/\sqrt{2} \tag{55}$$

and

$$\tilde{\varphi}_i(x) = [\varphi_{ia}(x) - i\varphi_{ib}(x)]/\sqrt{2}\,. \tag{56}$$

We require φ_{ia} and φ_{ib} to embody the same boundary conditions, so that the expansion of φ_{ia} and φ_{ib} utilizes the same constants, c_{ij} and $\tilde{c}_{ij}$. Consequently

$$\varphi_i(x) = \int d^3k[(c_{i1}A_{+1k} + c_{i2}A_{+2k})f_k + (\tilde{c}_{i1}A^\dagger_{-1k} + \tilde{c}_{i2}A^\dagger_{-2k})f^*_k]\,, \tag{57}$$

$$\tilde{\varphi}_i(x) = \int d^3k[(c_{i1}A_{-1k} + c_{i2}A_{-2k})f_k + (\tilde{c}_{i1}A^\dagger_{+1k} + \tilde{c}_{i2}A^\dagger_{+2k})f^*_k]\,, \tag{58}$$

where φ_i and $\tilde{\varphi}_i$ are related via the charge conjugation operator ([9]). Following the quantization pattern discussed above, with only minor changes due to the presence of two types of Fourier coefficient operators: positive charge, A_{+ik}, and negative charge, A_{-ik}, leads eventually to the following Green's function:

$$G_{22}(x-y) = G_F(x-y)\sin^2\theta_2 + G^*_F(x-y)\cos^2\theta_2\,. \tag{59}$$

Note that it has the same form as eq. (1). (Lagrangian interaction terms are expressed solely in terms of φ_2 and $\tilde{\varphi}_2$.) As a result we require $\gamma\tilde{\varphi}_2\gamma^{-1} = \varphi_2^\dagger$, the Hermitian conjugate of φ_2, where γ is the metric operator so that

$$C\varphi_2C^{-1} = \gamma\varphi_2^\dagger\gamma^{-1}\,, \tag{60}$$

where C is the charge conjugation operator.

3. – Fermion quantization.

In this section we describe our formulation of spin–one-half fermion quantum field theory. Again we are motivated by the need for a unique particle interpretation in nonstatic space-time. The formulation for fermions has close similarities to boson quantization.

Two fields are needed to describe a spin–one-half particle. The Lagrangian is

$$\mathscr{L} = \tilde{\psi}_2\gamma^0(i\nabla - m)\psi_1 + \tilde{\psi}_1\gamma^0(i\nabla - m)\psi_2 - \tilde{\psi}_1\gamma^0(i\nabla - m)\psi_1 . \tag{61}$$

We follow the conventions and notation of ref. (9). The fields $\tilde{\psi}_i$ will be related to the transpose of the charge conjugate field via

$$\tilde{\psi}_i = \psi_i^{cT}\gamma^0 C^T \tag{62}$$

for $i = 1, 2$. (In the usual formulation $\psi^\dagger = \tilde{\psi}$ would hold.) The equations of motion are

$$(i\nabla - m)\psi_i = 0 , \qquad \tilde{\psi}_i(i\overleftarrow{\nabla} - m) = 0 , \tag{63}$$

for $i = 1, 2$. The momentum conjugate to ψ_1 is

$$\pi_1 = i(\tilde{\psi}_2 - \tilde{\psi}_1) \tag{64}$$

and the conjugate to ψ_2 is

$$\pi_2 = i\tilde{\psi}_1 . \tag{65}$$

The canonical equal-time anticommutation relations imply

$$\{\psi_{1\alpha}(x), \tilde{\psi}_{1\beta}(y)\} = 0 , \tag{66}$$

$$\{\psi_{1\alpha}(x), \tilde{\psi}_{2\beta}(y)\} = \delta_{\alpha\beta}\delta^3(\boldsymbol{x} - \boldsymbol{y}) \tag{67}$$

and

$$\{\psi_{2\alpha}(x), \tilde{\psi}_{2\beta}(y)\} = \delta_{\alpha\beta}\delta^3(\boldsymbol{x} - \boldsymbol{y}) . \tag{68}$$

The most general form for the mode of expansion of the fields is (9)

$$\psi_i = \sqrt{2m}\sum_s \int \mathrm{d}^3k\,[(c_{i1}b_{1ks} + c_{i2}b_{2ks})f_k(x)u_{ks} + (\tilde{c}_{i1}d^\dagger_{1ks} + \tilde{c}_{i2}d^\dagger_{2ks})f^*_k(x)v_{ks}] . \tag{69}$$

Just as in the charged scalar case, we develop our formulation in such a way that the even and odd charge conjugation combinations, $\psi_i \pm \psi_i^c$, implement the same boundary conditions. Therefore

$$(70)\qquad \tilde{\psi}_i = \sqrt{2m}\sum_s \int d^3k[(c_{i1}d_{1ks} + c_{i2}d_{2ks})f_k(x)v^\dagger_{ks} + (\tilde{c}_{i1}b^\dagger_{1ks} + \tilde{c}_{i2}b^\dagger_{2ks})f^*_k(x)u^\dagger_{ks}]\,.$$

The nonzero Fourier-coefficient anti-commutation relations are

$$(71)\qquad \{d_{iks}, d^\dagger_{jk's'}\} = \{b_{iks}, b^\dagger_{jk's'}\} = (1-\delta_{ij})\,\delta_{ss'}\delta^3(\boldsymbol{k}-\boldsymbol{k}')$$

for i, $j = 1, 2$. The definition of states and inner products mirror the boson case. The vacua are defined by

$$(72)\qquad b_{1ks}|0\rangle_2 = b^\dagger_{1ks}|0\rangle_2 = d_{1ks}|0\rangle_2 = d^\dagger_{1ks}|0\rangle_2 = 0$$

and

$$(73)\qquad b_{2ks}|0\rangle_1 = b^\dagger_{2ks}|0\rangle_1 = d_{2ks}|0\rangle_1 = d^\dagger_{2ks}|0\rangle_1 = 0$$

and are related by a metric operator η:

$$(74)\qquad \eta|0\rangle_1 = |0\rangle_2$$

which satisfies $\eta = \eta^\dagger = \eta^{-1}$. We conventionally choose to construct kets from $|0\rangle_2$ and define their dual as their Hermitian conjugate multiplied by the metric operator. Thus inner products have the form

$$(75)\qquad \langle\alpha|\beta\rangle = (|\alpha\rangle)^\dagger\eta|\beta\rangle\,.$$

Physical observables must be self-adjoint, $A = A^* = \eta^{-1}A^\dagger\eta$, in order to have real eigenvalues. The Hamiltonian must be self-adjoint in order to conserve the norm. In view of eq. (68) we only use ψ_2 and $\tilde{\psi}_2$ in interaction terms and therefore require $\tilde{\psi}_2 = \eta^{-1}\psi^\dagger_2\eta$, so that

$$(76)\qquad C\psi_2C^{-1} = \eta^{-1}C\tilde{\psi}^T_2\eta\,,$$

which bears comparison with eq. (15.112) of ref. (9) and also eq. (60) above. The equal-time anticommutation relations, eq. (76), and the adjointness of H restrict the constants c_{ij} and $\tilde{c}_{ij}$ so that

$$(77)\qquad \psi_1 = \sqrt{2m}\sum_s \int d^3k[(\cos(\theta_1-\theta_2)b_{1ks} + \sin(\theta_1-\theta_2)b_{2ks})f_k(x)u_{ks}/\sin\theta_1 + \\ + (\cos(\theta_1-\theta_2)d^\dagger_{1ks} - \sin(\theta_1-\theta_2)d^\dagger_{2ks})f^*_k(x)v_{ks}/\cos\theta_1]$$

and

$$(78)\qquad \psi_2 = \sqrt{2m}\sum_s \int d^3k[(\sin\theta_2 b_{1ks} + \cos\theta_2 b_{2ks})f_k(x)u_{ks} + \\ + (\cos\theta_2 d^\dagger_{1ks} + \sin\theta_2 d^\dagger_{2ks})f^*_k(x)v_{ks}]\,.$$

The Hamiltonian is

$$H = \sum_s \int d^3k\,\omega_k (b^\dagger_{2ks} b_{1ks} - b_{2ks} b^\dagger_{1ks} + d^\dagger_{2ks} d_{1ks} - d_{2ks} d^\dagger_{1ks}) \,. \tag{79}$$

In contrast to the usual formulation, we see that our Hamiltonian does not have an infinite vacuum energy with respect to $|0\rangle_2$. It is not positive definite, but we will be able to develop a unitary S-matrix theory in the space of positive-energy asymptotic states, if we choose $\theta_2 = \pi/2$. This is evident from an examination of the Green's function

$$S_{22}(x-y) = -i\,{}_1\langle 0|T(\psi_2(x)\,\bar{\psi}_2(y)\gamma_0)|0\rangle_2 = \tag{80}$$

$$= \sin^2\theta_2 S_F(x-y) + \cos^2\theta_2 C\gamma^0 S^*_F(C\gamma^0)^{-1} \,, \tag{81}$$

which gives $S_{22} = S_F$, the usual Feynman propagator, if $\theta_2 = \pi/2$. As in the boson case, we introduce interactions only through type-2 operators, and use type-2 operators to LSZ reduce in and out particles. The result is a perturbation theory which, in the Fermion sector, only involves time-ordered products of ψ_2 and $\bar{\psi}_2$. Thus we can establish a model in our formulation which is equivalent, so far as S-matrix elements are concerned, to any given model of the usual formulation. In particular, our model electrodynamics has the Lagrangian

$$\mathscr{L} = \tfrac{1}{2} F^1_{\mu\nu} F^{2\mu\nu} + \tfrac{1}{4} F^1_{\mu\nu} F^{1\mu\nu} + \bar{\psi}_2\gamma^0(i\nabla - m)\psi_1 + \bar{\psi}_1\gamma^0(i\nabla - m)\psi_2 - \bar{\psi}_1\gamma^0(i\nabla - m)\psi_1 - e_0\bar{\psi}_2\gamma^0\hat{A}_2\psi_2 \tag{82}$$

with

$$F^i_{\mu\nu} = \partial_\nu A_{i\mu} - \partial_\mu A_{i\nu} \tag{83}$$

for $i = 1, 2$. While we will discuss this model more fully elsewhere ([6]) two things are worth noting. First the interaction is expressed solely in terms of fields of type 2—both for fermions and the photon. Following our quantization procedure leads to S-matrix expressions which are term-by-term equal to corresponding expressions in QED. Secondly, the model is gauge invariant. The gauge transformation is

$$\psi_2 \to \exp[i\Lambda]\psi_2 \,, \tag{84}$$

$$\bar{\psi}_2 \to \exp[-i\Lambda]\bar{\psi}_2 \,, \tag{85}$$

$$A_{2\mu} \to A_{2\mu} - \frac{1}{e}\partial_\mu\Lambda \,, \tag{86}$$

$$\psi_1 \to \psi_1 + (\exp[i\Lambda] - 1)\psi_2 \,, \tag{87}$$

$$\bar{\psi}_1 \to \bar{\psi}_1 + (\exp[-i\Lambda] - 1)\bar{\psi}_2 \,, \tag{88}$$

and its associated conserved current is

$$J_\mu = -i\frac{\delta\mathscr{L}}{\delta\partial^\mu\psi_1}\psi_2 - i\frac{\delta\mathscr{L}}{\delta\partial^\mu\psi_2}\psi_2\,, \tag{89}$$

$$J_\mu = \bar{\psi}_2\gamma^0\gamma_\mu\psi_2\,. \tag{90}$$

We close our discussion of fermions by considering the case $\theta_2 = \pi/4$ which, by eq. (81), gives the principal-value propagator

$$S_{22}(x-y) = \int\frac{\mathrm{d}^4k}{(2\pi)^4}\exp[-ik\cdot(x-y)](k+m)\frac{P}{k^2-m^2}\,, \tag{91}$$

where

$$\frac{P}{k^2-m^2} = \frac{1}{2}\left(\frac{1}{k^2-m^2+i\varepsilon}+\frac{1}{k^2-m^2-i\varepsilon}\right). \tag{92}$$

Thus a quantum action-at-a-distance model of fermions can be constructed within the framework of our formulation.

4. – Particle interpretation.

In this section we shall show that the particle interpretation of our formulation of quantum field theory is well defined for the case of a free scalar particle in a nonstatic space-time where no global timelike co-ordinate exists. We assume an action of the form

$$S = \int\mathrm{d}^4x\,[\varphi_2 D\varphi_1 - \tfrac{1}{2}\varphi_1 D\varphi_1] + (\text{surface terms})\,, \tag{93}$$

which under the variation of S gives the field equations

$$D\varphi_1 = D\varphi_2 = 0\,. \tag{94}$$

The self-adjointness of D implies

$$\int_V [f^* Dg - (Df)^* g]\,\mathrm{d}^4x = \int_{\Sigma_v} f^*\overleftrightarrow{D}^\mu g\,\mathrm{d}\Sigma_\mu\,, \tag{95}$$

where Σ_v is the surface bounding V, $\mathrm{d}\Sigma_\mu$ is an outward directed surface element of Σ_v and D^μ is a two-edged vector differential operator. If Σ is a spacelike complete Cauchy hypersurface for the field equations (we assume they exist), then an inner product for complex solutions of the field equations can be defined by

$$(v_1, v_2) = i\int_\Sigma v_1^* \overleftrightarrow{D}^\mu v_2 \, d\Sigma_\mu . \tag{96}$$

We now choose an arbitrary complete orthonormal set of pairs of complex conjugate solutions of eq. (94), $\{V_\alpha, V_\alpha^*\}$, satisfying

$$(V_\alpha, V_{\alpha'}) = -(V_\alpha^*, V_{\alpha'}^*) = \delta_{\alpha\alpha'} , \tag{97}$$

$$(V_\alpha, V_{\alpha'}^*) = 0 , \tag{98}$$

and use them in the mode expansion of the field operators

$$\varphi_i = \sum_\alpha [(c_{i1} A_{1\alpha} + c_{i2} A_{2\alpha}) V_\alpha + (\tilde{c}_{i1} A_{1\alpha}^\dagger + \tilde{c}_{i2} A_{2\alpha}^\dagger) V_\alpha^*] , \tag{99}$$

where c_{ij} and $\tilde{c}_{ij}$ are real c-numbers. The Fourier coefficient operators satisfy

$$[A_{i\alpha}, A_{j\alpha'}^\dagger] = (1 - \delta_{ij}) \delta_{\alpha\alpha'} \tag{100}$$

with all other commutators equal to zero. The commutativity of Fourier-coefficient operators and their Hermitian conjugate allows us to define the vacua $|0\rangle_1$ and $|0\rangle_2$ by

$$A_{1\alpha}|0\rangle_2 = A_{1\alpha}^\dagger|0\rangle_2 = A_{2\alpha}|0\rangle_1 = A_{2\alpha}^\dagger|0\rangle_1 = 0 . \tag{101}$$

We choose to construct states from $|0\rangle_2$. The one-particle ket corresponding to the Fourier transform variable α is

$$|\alpha\rangle = -(v_\alpha^*, \varphi_2)|0\rangle_2/\tilde{c}_{22} \tag{102}$$

and the one-particle bra dual to it is

$$\langle\alpha| = (|\alpha\rangle)^\dagger \gamma , \tag{103}$$

where γ is a metric operator satisfying $\gamma = \gamma^{-1} = \gamma^\dagger$ and

$$\gamma^{-1} A_{i\alpha} \gamma = \varepsilon_{ij} A_{j\alpha} \tag{104}$$

with $\varepsilon_{11} = \varepsilon_{22} = 0$ and $\varepsilon_{12} = \varepsilon_{21} = 1$. The further development of this quantization proceeds along the lines of sect. **2**. In particular φ_2 is self-adjoint.

We now introduce a quantization of the particle described by S which parallels the above development in every detail except that a different complete orthonormal set of field equation solutions $\{W_\beta, W_\beta^*\}$ is used in the mode expansion of the fields

$$\varphi_i = \sum_\beta [(c_{i1} A_{1\beta} + c_{i2} A_{2\beta}) W_\beta + (\tilde{c}_{i1} A^\dagger_{1\beta} + \tilde{c}_{i2} A^\dagger_{2\beta}) W^*_\beta]. \tag{105}$$

The question arises: how are we to relate the two quantizations? In the usual formulation only one answer is apparent—the field operators are to be identified [1,2], since they are uniquely determined by the field equations and the canonical commutation relations. But in the present case, the field operators are not uniquely determined, so that the identification of the fields in eq. (105) and (99) is not required. The relation between the quantizations must obviously be well defined (in the sense that every operator and state in one quantization can be uniquely expressed in terms of operators and states of the other representations). More importantly, it must only relate operators whose properties are fixed by the field equation and the canonical commutation relations; and whose expectation values are uniquely specified by purely geometrical restrictions on their support and do not embody a definition of positive frequency. In our formalism, the operators which satisfy these requirements are linear combinations of

$$\varphi^{\text{H}}_{iv} = \sum_\alpha (A_{i\alpha} V_\alpha + A^\dagger_{i\alpha} V^*_\alpha), \tag{106}$$

$$\varphi^{\text{H}}_{iw} = \sum_\beta (A_{i\beta} W_\beta + A^\dagger_{i\beta} W^*_\beta), \tag{107}$$

for $i = 1, 2$ in the respective quantizations we can restrict the discussion to these quantities. In particular, the vacuum expectation value

$$_1\langle 0|\varphi^{\text{H}}_1(x)\varphi^{\text{H}}_2(y)|0\rangle_2 = \tfrac{1}{2}\,{}_1\langle 0|[\varphi^{\text{H}}_1(x), \varphi^{\text{H}}_2(y)]|0\rangle_2, \tag{108}$$

$$_1\langle 0|\varphi^{\text{H}}_1(x)\varphi^{\text{H}}_2(y)|0\rangle_2 = \frac{i}{2}\Delta(x-y), \tag{109}$$

where $\Delta(x-y)$, the commutator function, has the well-defined geometrical property that it vanishes at spacelike distances.

Identifying φ^{H}_{iw} with φ^{H}_{iv}, for $i = 1, 2$, leads to the relations

$$A_{i\beta} = (W_\beta, \varphi^{\text{H}}_i) = \sum_\alpha [(W_\beta, V_\alpha) A_{i\alpha} + (W_\beta, V^*_\alpha) A^\dagger_{i\alpha}], \tag{110}$$

for $i = 1, 2$ plus Hermitian-conjugate expressions. The form of the inner products on the right-hand side of eq. (110) is determined by requiring that the definition of positive frequency implicit in the separation of the orthonormal set $\{W_\beta, W^*_\beta\}$ into complex conjugate pairs of solutions can also be implemented by linear combinations of V_α and V^*_α. Specifically, we assume that a complete orthonormal set of pairs of complex conjugate functions, $\{V_\alpha, V^*_\alpha\}$, exists which satisfies

$$V_\alpha = c_1 \tilde{V}_\alpha + c_2 \tilde{V}^*_\alpha, \tag{111}$$

$$V^*_\alpha = c^*_1 \tilde{V}^*_\alpha + c^*_2 \tilde{V}_\alpha, \tag{112}$$

for all α, where c_1 and c_2 are c-number functions of α only with $|c_1| > |c_2|$, and where

$$(W_\beta, \tilde{V}^*_\alpha) = 0, \tag{113}$$

$$(W^*_\beta, \tilde{V}_\alpha) = 0, \tag{114}$$

for all β. The orthogonality conditions imply

$$c_1 = \exp[i\lambda_1]\cosh\lambda_2, \tag{115}$$

$$c_2 = \exp[i\lambda_1]\sinh\lambda_2, \tag{116}$$

where λ_1 and λ_2 are solely functions of α. The substitution of eqs. (111) and (112) in eq. (110) and use of eqs. (113) and (114) gives

$$A_{i\beta} = \sum_\alpha (W_\beta, \tilde{V}_\alpha)[\exp[i\lambda_1]\cosh\lambda_2 A_{i\alpha} + \exp[-i\lambda_1]\sinh\lambda_2 A^\dagger_{i\alpha}] \tag{117}$$

for $i = 1, 2$. Note that the bracketed term on the right-hand side of the equation has the same form as the Bogoliubov rotated Fourier-coefficient operator given in eq. (23). In the present case we can rewrite eq. (117) in the form

$$A_{i\beta} = \sum_\alpha (W_\beta, \tilde{V}_\alpha) B_{\lambda_1\lambda_2} A_{i\alpha} B^{-1}_{\lambda_1\lambda_2} \tag{118}$$

with

$$B_{\lambda_1\lambda_2} = \exp\left[2i\sum_\alpha \lambda_1(\alpha)\Gamma_{3\alpha}\right]\exp\left[2i\sum_\alpha \lambda_2(\alpha)\Gamma_{2\alpha}\right], \tag{119}$$

where $\Gamma_{3\alpha}$ and $\Gamma_{2\alpha}$ are obtained from eqs. (25) and (26) by replacing the subscripts k with α.

The particle interpretations of the two quantizations will now be shown to be identical. First we note that the vacuum $|0\rangle_2$ of the «α» quantization is invariant under $B_{\lambda_1\lambda_2}$, so that it may be taken to be identical with the $|0\rangle_2$ vacumm of the «β» quantization. Next we note that the canonical commutation relations and the vacuum expectation value of any product of field operators are invariant under B:

$${}_1\langle 0|\varphi_{i_1}(x_1)\varphi_{i_2}(x_2)\ldots|0\rangle_2 = {}_1\langle 0|B^{-1}\varphi_{i_1}(x_1)\varphi_{i_2}(x_2)\ldots B|0\rangle_2. \tag{120}$$

This implies that we could replace $A_{i\alpha}$ with $B_{\lambda_1\lambda_2}A_{i\alpha}B^{-1}_{\lambda_1\lambda_2}$ in the mode expansions, eq. (107), with no change in physical consequences. In particular, this applies to the definition of particle kets. Equation (102) becomes

$$|\alpha\rangle = (\exp[-i\lambda_1]\cosh\lambda_2 A^{\dagger}_{2\alpha} + \exp[i\lambda_1]\sinh\lambda_2 A_{2\alpha})|0\rangle_2 . \tag{121}$$

Consequently, $A^{\dagger}_{2\beta}|0\rangle_2$ is a superposition of one-particle states in the « α » quantization. In general, the N-particle state in the « β » quantization is a superposition of N-particle states in the « α » quantization.

The invariance of particle number under Bogoliubov transformations is reflected in the relation between the particle number operator,

$$N = \sum_{\alpha}(A^{\dagger}_{2\alpha}A_{1\alpha} - A_{2\alpha}A^{\dagger}_{1\alpha}) , \tag{122}$$

which is invariant under Bogoliubov transformations, and related to the Casimir operator of the Bogoliubov group (cf. eqs. (29)-(31)). Our identification of N as the particle number operator is based, as most charge and number operators are, on an invariance of the action under a global change of phase of fields. In our case we note that the action of eq. (93) is invariant under the infinitesimal phase change

$$\varphi_1 \to \varphi_1 + i\varepsilon\varphi_1 , \qquad \varphi_2 \to \varphi_2 + i\varepsilon(\varphi_1 - \varphi_2) . \tag{123}$$

The corresponding conserved-number operator is given by

$$N = i\int_{\Sigma_\nu}\varphi_1 \overleftrightarrow{D}^{\mu}\varphi_2 \, d\Sigma_\mu . \tag{124}$$

Because φ_1 and φ_2 implicitly embody a definition of positive frequency, we are led to replace them with Hermitian operators:

$$N = i\int_{\Sigma_\nu}\varphi_1^{\text{H}} D^{\mu}\varphi_2^{\text{H}} \, d\Sigma_\mu \tag{125}$$

with φ_i^{H} given in eq. (106). Equation (125) can be evaluated by using eq. (96) and (97) to give eq. (122). Thus our definition of number operator is physically motivated. It is also consistent with our expectations of a number operator.

We shall now summarize our picture of second quantization in curved space-time where no global timelike Killing vector is present. Consider a complete spacelike Cauchy hypersurface. At each point on the surface there is a local timelike direction. There is, in general, a class of operators, which will locally generate a displacement in the timelike direction, but which globally generate very different motions. Due to the absence of a global timelike Killing vector, no member of the class of potential generators of the dynamics is physically selected as the generator of the dynamics. One is free to choose any member as the generator of the dynamics locally. Each choice implies a different definition of positive and negative frequency when field operators are represented by Fourier expansions.

In the usual formulation of quantum field theory any choice of generator of the dynamics (and thus Fourier representation of field operators) is unitarily inequivalent to any other choice in general. As a result each choice gives a *different physical theory* and the second quantization of a theory is not unique. Practically, this means that 1) a one-particle state in one quantization is a many-particle state in any other quantization (particle number is ambiguous), 2) in general one can construct a one-particle state which is an eigenstate of a generator in one representation, but one cannot construct a one-particle state in another representation which is an eigenstate of the same generator (the space of states is different), and 3) (if interactions are introduced) the S-matrix differs from quantization to quantization. Obviously, there are only two acceptable alternatives in this situation; either some new principle selects one representation as the correct physical representation, or a modification of quantum field theory is necessary. In the absence of a new physical principle, we have formulated a modification of quantum field theory.

Our formulation allows one to quantize a field theory with any of the potential generators of the dynamics and yet to have a physically unique theory. Different quantizations can be related by Bogoliubov transformations and, in our formulation, are unitarily equivalent. Consequently, particle number is invariant—N-particle states in one representation are superpositions of N-particle states in any other representation; the set of states in one representation is unitarily equivalent to the set of states in any other representation; and the S-matrix is uniquely determined in the case of an interacting theory (the proof is analogous to that of the flat space-time case discussed in sect. 2). Our formulation associates a unique physical theory with any given action. In a sense, it implements an equivalence principle in the space of solutions to the field equations—any complete orthonormal set of solutions to the field equations can be used in the Fourier expansion of field operators and a unique physical theory results (cf. eqs. (111)-(114)).

In conclusion, we note that the problem we have addressed relates to one observer and the ambiguities of conventional quantum field theory he must face. Different observers in relatively accelerating frames will not see the same number of particles in our formulation. Neither is particle creation near black holes precluded in our formulation.

* * *

I am grateful to the Aspen Center for Physics for its hospitality while part of this work was being done, and to M. A. B. BEG, S. MANDELSTAM and D. PARK for stimulating conversations.

● RIASSUNTO (*)

Si descrive una generalizzazione della teoria quantistica dei campi che ha un'unica interpretazione particellare — anche negli spazio-tempo in cui non esiste alcuna coordinata globale di tipo tempo. La formulazione è descritta in dettaglio per i casi di bosoni scalari e di fermioni con spin ½ nello spazio-tempo piatto. Si mostra che è possibile costruire un modello nel nostro approccio che è fisicamente equivalente a un qualsiasi modello nella solita formulazione. Inoltre, si può costruire una nuova classe di modelli che non sono possibili nella solita formulazione. Questa classe comprende modelli quantici di azione a distanza che possono essere usati per sviluppare modelli con equazioni di campo a derivata più alta che sono unitarie. La nostra formulazione permette ampia scelta delle condizioni limite, cosicché si può optare per un continuo di possibili funzioni di Green che vanno dai propagatori di Feynman a propagatori del valore principale (mezzo avanzato-mezzo ritardato).

(*) *Traduzione a cura della Redazione.*

Локальное определение асимптотических состояний частиц.

Резюме (*). — Описывается обобщение квантовой теории поля, которая имеет единую частичную интерпретацию — даже в пространстве и времени, где не существует глобальной времениподобной координаты. Подробно описывается формулировка для случаев скалярных бозонов и фермионов со спином половина в плоском пространстве-времени. Мы показываем, что имеется возможность сконструировать модель в нашем подходе, которая физически эквивалентна любой заданной модели в обычной формулировке. Кроме того, может быть сконструирован новый класс моделей, которые являются невозможными в обычной формулировке. Этот класс включает модели квантового действия на расстоянии, которые могут быть использованы для развития моделей с полевыми уравнениями с высшими производными, которые являются унитарными. Наша формулировка допускает некоторую свободу в выборе граничных условий, так что имеется возможность выбрать континуум возможных гриновских функций, от фейнмановских пропагаторов до пропагаторов главных значений (наполовину опережающая - наполовину запаздывающая).

(*) *Переведено редакцией.*

Appendix B. Non-Abelian Quantum Field Theories in Non-Static Space-times

This refereed paper is S. Blaha, Il Nuovo Cimento **49 A**, 113 (1979). Reprinted with the kind permission of Il Nuovo Cimento. It extends the discussion of principal value propagators of quantum field theories, particularly those with higher order derivatives, to support the local definition of particle states using non-static non-rectangular and curved space-time coordinates. Principal values propagators were used in S. Blaha, Phys. Rev. D**10**, 4268 (July, 1974) and Phys. Rev. D**11**, 2921 (1974) in the definition of higher order derivative quantum field theories for quark confinement with an explicit linear potential. Thus the author's strong interaction quantum field theories are now shown to have a unique particle interpretation in both the usual rectangular coordinate systems and other non-static coordinate systems in both flat and curved space-times.

IL NUOVO CIMENTO VOL. 49 A, N. 1 1 Gennaio 1979

New Framework for Gauge Field Theories (*).

S. BLAHA
Physics Department, Williams College - Williamstown, Ma. 01267

(ricevuto l'8 Agosto 1978)

Summary. — We formulate gauge theories within the framework of a generalization of quantum field theory. In particular, we discuss models of electrodynamics and of Yang-Mills theories, a model of the strong interaction with higher-order derivatives and quark confinement and a renormalizable model of pure quantum gravity with Einstein Lagrangian. In the case of electrodynamics we show that two models are possible: one with predictions which are identical to QED and one which is a quantum action-at-a-distance model of electrodynamics. In the case of Yang-Mills theories we can construct a model which is identical in predictions to any conventional model, or a quantum action-at-a-distance model. In the second case it is possible to eliminate all loops of Yang-Mills particles (in all gauges) in a manner consistent with unitarity. A variation of Yang-Mills models exists in our formulation which has higher-order derivative field equations. It is unitary and has positive probabilities. It can be used to construct a model of the strong interactions which has a linear potential and manifest quark confinement. Finally we show how to construct an action-at-a-distance model of pure quantum gravity (whose classical limit is the dynamics of the Einstein Lagrangian) coupled to an external classical source. The model is trivially renormalizable.

1. – Introduction.

Because of the absence of an acceptable physical interpretation of conventional quantum field theory in the case of curved nonstatic space-time, we

(*) Supported in part by Research Corporation and the National Science Foundation.

recently developed a modified formulation of quantum field theory (1). We showed it had a unique physical particle interpretation in nonstatic space-time. We described the flat-space-time formulation for the cases of scalar particles and spin-one-half fermions. In both cases we found that it was possible to construct a model which was equivalent in predictions to any model of the usual formulation. However, it was also possible to construct models which had no analogue in the usual formulation. In these models the quantum exchange of a « particle » did not take place via a Feynman propagator. Rather, the propagator G could be chosen to be any of a continuum of possibilities ranging from the Feynman propagator to the principal-value (half advanced-half retarded) propagator. The choice is parametrized by an angle θ, whose specification is equivalent to a choice of boundary condition

$$G(x-y) = \sin^2\theta\, G_F(x-y) + \cos^2\theta\, CG_F^* C^{-1}, \tag{1}$$

where G_F is the usual Feynman propagator with G_F^* its complex conjugate and C is an appropriate charge conjugation matrix.

The new degree of freedom, represented by θ, leads to new possibilities for the formulation of flat-space-time quantum field theory models which will be especially evident in the case of gauge theories.

In sect. **2** we shall explore models of quantum electrodynamics which occur within the framework of our formulation. We shall show that one model is completely equivalent to QED in its predictions. In addition, we shall show that a quantum action-at-a-distance electrodynamics is also possible.

In sect. **3** we describe model Yang-Mills theories and show that a model equivalent to any conventional model can be formulated as well as a quantum action-at-a-distance model. In the second case all non-Abelian boson loops can be eliminated (in all gauges) in a manner which is consistent with unitarity.

In sect. **4** we describe a non-Abelian model of quark confinement based on higher-derivative field equations for which a unitary physical S-matrix is obtained by the choice of principal-value propagators. This demonstrates the utility of principal-value propagators for widening the class of unitary quantum field theories to include higher-derivative theories.

In sect. **5** we describe a model of pure quantum gravity based on the Einstein Lagrangian which is trivially renormalizable, if gravitons propagate via principal-value propagators. (No graviton loops.) This illustrates the potential of the principal-value propagator to ameliorate renormalization problems by allowing one to limit the self-interactions of quantum fields.

In view of the unfamiliar features of principal-value propagators, we shall

(1) S. BLAHA: *Nuovo Cimento*, **49** A, 35 (1978).

devote the remainder of this section to a discussion of their properties in relation to causality and analyticity.

Among the first appearances of principal-value, or half advanced-half retarded, propagators was in Feynman's space-time approach to quantum electrodynamics ([2]), where he observed that one is naturally led to such a propagator for the photon, if one considers a quantum one-photon exchange between two electrons wherein the photons propagate, as in classical electrodynamics, via retarded propagators. Since one cannot distinguish between the process where a photon propagates from electron A to B and the process where the photon propagates from B to A for short time intervals one must sum the two amplitudes and a principal-value propagator results. This observation illustrates the absence of a direct relation between the propagator of the classical field and the propagator of the corresponding quantum field. In particular, it is perfectly possible for a quantum field to have a principal-value propagator, and yet have the classical field propagate via a retarded propagator. This can be understood within the framework of the absorber model of Feynman and Wheeler ([3]). For a quantum process, where a finite number of quanta is exchanged, the effective propagator of the quanta cannot have this aspect of its character changed. But in a classical situation, where infinite numbers of quanta are involved, the effective propagator of the quanta can be changed through interaction with the absorber in the manner outlined in ref. ([3]). We therefore conclude that the propagation of quanta via principal-value propagators does not imply that macroscopic causality (as represented by retarded propagators) is lost. This remark is relevant to our models of the strong interaction and quantum gravity discussed later.

So far as microscopic causality is concerned, it will be seen that principal-value propagators are completely consistent with the vanishing of commutators of field operators for spacelike distances. Thus neither macroscopic nor microscopic causality is necessarily inconsistent with the use of quantum principal-value propagators.

The use of principal-value propagators does lead to a different analytic structure of S-matrix amplitudes. Amplitudes are now piecewise analytic. However, several authors ([4]) have shown that such amplitudes are not necessarily inconsistent with experiment—even for the case of the pion-nucleon dispersion relations. In particular, considering the absence of any direct relation between the analytic properties of quark-quark scattering amplitudes,

---

([2]) R. FEYNMAN: *Phys. Rev.* **76**, 769 (1949), footnote ([7]).

([3]) J. WHEELER and R. FEYNMAN: *Rev. Mod. Phys.*, **17**, 157 (1945); **21**, 425 (1949); F. HOYLE and J. NARLIKAR: *Ann. of Phys.*, **54**, 207 (1969); **62**, 44 (1971).

([4]) R. E. CUTKOWSKY, P. V. LANDSHOFF, D. I. OLIVE and J. C. POLKINGHORNE: *Nucl. Phys.*, **12** B, 281 (1969); T. D. LEE and G. C. WICK: *Phys. Rev. D*, **2**, 1033 (1970); M. GUNDZIK and E. C. G. SUDARSHAN: *Phys. Rev. D*, **6**, 798 (1972); H. RECHENBERG and E. C. G. SUDARSHAN: *Nuovo Cimento*, **14** A, 299 (1973).

and the analytic properties of the scattering amplitudes of their bound states one cannot rule out the possibility of principal-value propagators for color gluons. In the case of quantum gravity, another potential application of principal-value propagators, the analytic structure of scattering amplitudes due to graviton exchange is unknown and likely to remain so. In the absence of such information, the renormalizability of pure quantum gravity, and the compatibility with Mach's principle, resulting from the use of principal-value propagators for gravitons is encouraging.

Perhaps the most important question facing field theory models with principal-value propagators is unitarity. The appendix contains a detailed discussion of this issue. The physical S-matrix is shown to be unitary. In addition the value of using infinite-momentum frame variables to compute S-matrix elements is pointed out.

2. – Model electrodynamics.

In this section we shall describe a model electrodynamics which is identical to quantum electrodynamics in its predictions. We shall also discuss a model for quantum action-at-a-distance electrodynamics.

The Lagrangian density of our models [1] is

$$(2)\qquad \mathscr{L} = -\tfrac{1}{2}F^1_{\mu\nu}F^{2\mu\nu} - \tfrac{1}{4}F^1_{\mu\nu}F^{1\mu\nu} + \tilde{\psi}_2\gamma^0(i(\gamma\cdot\nabla) - e_0(\gamma\cdot A_2) - m)\psi_2 + (\tilde{\psi}_1 - \tilde{\psi}_2)\gamma^0(i(\gamma\cdot\nabla) - m)(\psi_1 - \psi_2)\,,$$

where we have introduced two electromagnetic fields, $A_{1\mu}$ and $A_{2\mu}$, so that

$$(3)\qquad F^i_{\mu\nu} = \partial_\nu A_{i\mu} - \partial_\mu A_{i\nu}\,,$$

for $i = 1, 2$, and two fields, ψ_1 and ψ_2 for electrons. The field $\tilde{\psi}_i$ is related to the transpose of the charge conjugate of ψ_i by [1]

$$(4)\qquad \tilde{\psi}_i = \psi_i^{cT}\gamma^0 C^T\,,$$

where C is a charge conjugation matrix. We follow the notation and conventions of ref. [5]. The free field theory for fermions has been developed in detail in ref. [1].

---

[5] J. BJORKEN and S. DRELL: *Relativistic Quantum Fields* (New York, N. Y., 1965).

Before discussing the free field theory of the photons we note the invariance of $\mathscr{L}$ under a restricted gauge transformation of the second kind,

$$\psi_2 \to \psi_2 \exp[i\Lambda], \tag{5}$$

$$\tilde{\psi}_2 \to \tilde{\psi}_2 \exp[-i\Lambda], \tag{6}$$

$$e_0 A_{2\mu} \to e_0 A_{2\mu} - \partial_\mu \Lambda, \tag{7}$$

$$\psi_1 \to \psi_1 + [\exp[i\Lambda] - 1]\psi_2, \tag{8}$$

$$\tilde{\psi}_1 \to \tilde{\psi}_1 + [\exp[-i\Lambda] - 1]\tilde{\psi}_2. \tag{9}$$

The associated conserved current is

$$J_\mu = \tilde{\psi}_2 \gamma^0 \gamma_\mu \psi_2. \tag{10}$$

It should be noted that the Lagrangian is also invariant under an independent gauge transformation, $A_{1\mu} \to A_{1\mu} - \partial_\mu \Lambda_1$.

Upon varying the action associated with $\mathscr{L}$, we find the field equations

$$\partial^\mu F^1_{\mu\nu} = \partial^\mu F^2_{\mu\nu}, \tag{11}$$

$$\partial^\nu F^2_{\mu\nu} = e_0 \tilde{\psi}_2 \gamma^0 \gamma_\nu \psi_2, \tag{12}$$

$$(i(\gamma\cdot\nabla) - m)\psi_1 = (i(\gamma\cdot\nabla) - m)\psi_2, \tag{13}$$

$$(i(\gamma\cdot\nabla) - e_0(\gamma\cdot A_2) - m)\psi_2 = 0. \tag{14}$$

Equations (12) and (14) demonstrate the equivalence of our model to the usual model of electrodynamics at the level of c-number fields.

We now turn to the case of free photons. From the Lagrangian we can identify the canonical momentum conjugate to $A_{1\mu}$ as

$$\pi_{1\mu} = F^2_{0\mu} - F^1_{0\mu}, \tag{15}$$

while the momentum conjugate to $A_{2\mu}$ is

$$\pi_{2\mu} = F^1_{0\mu}. \tag{16}$$

Since the free electromagnetic Lagrangian is invariant under independent gauge transformations of $A_{1\mu}$ and $A_{2\mu}$ the choice of gauges for $A_{1\mu}$ and $A_{2\mu}$ are also independent. We shall second quantize the fields in the joint Coulomb gauge

$$\vec{\nabla}\cdot\vec{A}_1 = \vec{\nabla}\cdot\vec{A}_2 = 0. \tag{17}$$

The resulting equal-time commutation relations are

$$[F^1_{0i}(\vec{x}, t), A_{1j}(\vec{y}, t)] = 0 , \tag{18}$$

$$[F^1_{0i}(\vec{x}, t), A_{2j}(\vec{y}, t)] = i\delta^{\text{tr}}_{ij}(\vec{x} - \vec{y}) , \tag{19}$$

$$[F^2_{0i}(\vec{x}, t), A_{2j}(\vec{y}, t)] = i\delta^{\text{tr}}_{ij}(\vec{x} - \vec{y}) , \tag{20}$$

$$[F^2_{0i}(\vec{x}, t), A_{1j}(\vec{y}, t)] = i\delta^{\text{tr}}_{ij}(\vec{x} - \vec{y}) , \tag{21}$$

with

$$\delta^{\text{tr}}_{ij}(\vec{x} - \vec{y}) = \int \frac{\mathrm{d}^3k}{(2\pi)^3} \exp[-i\vec{k}\cdot(\vec{x} - \vec{y})](\delta_{ij} - k_i k_j/|\vec{k}|^2) . \tag{22}$$

The field operator expansions are completely analogous to the expansion of a free scalar field discussed in ref. (1). The major difference is the necessary presence of polarization vectors in the electromagnetic-field expansions

$$\vec{A}_1(x) = \int \mathrm{d}^3k \sum_{\lambda=1}^{2} \vec{\varepsilon}(k, \lambda)[f_k(x)(\cos(\theta_1 - \theta_2)a_{1k\lambda} + \sin(\theta_1 - \theta_2)a_{2k\lambda})/\sin\theta_1 + \\ + f^*_k(x)(\cos(\theta_1 - \theta_2)a^\dagger_{1k\lambda} - \sin(\theta_1 - \theta_2)a^\dagger_{2k\lambda})/\cos\theta_1] \tag{23}$$

and

$$\vec{A}_2(x) = \int \mathrm{d}^3k \sum_{\lambda=1}^{2} \vec{\varepsilon}(k, \lambda)[f_k(x)(\cos\theta_2 a_{2k\lambda} + \sin\theta_2 a_{1k\lambda}) + \\ + f^*_k(x)(\sin\theta_2 a^\dagger_{2k\lambda} + \cos\theta_2 a^\dagger_{1k\lambda})] , \tag{24}$$

where θ_1 and θ_2 are arbitrary angles. As discussed in ref. (1) the choice of these angles represents a choice of boundary conditions for the free-field Green's functions. If the Fourier coefficient operators satisfy

$$[a_{ik\lambda}, a_{jk'\lambda'}] = [a^\dagger_{ik\lambda}, a^\dagger_{jk'\lambda'}] = 0 , \tag{25}$$

$$[a_{ik\lambda}, a^\dagger_{jk'\lambda'}] = (1 - \delta_{ij})\delta_{\lambda\lambda'}\delta^3(\vec{k} - \vec{k}') , \tag{26}$$

for $i, j = 1, 2$, then the equal-time commutation relations will be satisfied for arbitrary θ_1 and θ_2. In addition the Hamiltonian which is the 0-0 component of the energy-momentum tensor derived from the Lagrangian, has the form

$$H = \int \mathrm{d}^3k \sum_{\lambda=1}^{2} \omega_k(a^\dagger_{2k\lambda}a_{1k\lambda} + a_{2k\lambda}a^\dagger_{1k\lambda}) , \tag{27}$$

independent of θ_1 and θ_2. There is an analogy (6) between the quantization

(6) S. BLAHA: *Phys. Rev. D*, **17**, 994 (1978).

of the mode amplitudes, eqs. (25) and (26), and the co-ordinates of an assembly of one-dimensional harmonic oscillators just as in the second quantization of conventional field theory. We shall define photon vacua in the present case in a manner consistent with the procedure of ref. (1) and in analogy with the simple harmonic-oscillator model of ref. (6):

$$a_{1k\lambda}|0\rangle_2 = a^\dagger_{1k\lambda}|0\rangle_2 = 0\,, \tag{28}$$

$$a_{2k\lambda}|0\rangle_2 \neq 0\,, \quad a^\dagger_{2k\lambda}|0\rangle_2 \neq 0\,, \tag{29}$$

for one vacuum, $|0\rangle_2$, and

$$a_{2k\lambda}|0\rangle_1 = a^\dagger_{2k\lambda}|0\rangle_1 = 0\,, \tag{30}$$

$$a_{1k\lambda}|0\rangle_1 \neq 0\,, \quad a^\dagger_{1k\lambda}|0\rangle_1 \neq 0\,, \tag{31}$$

for the other vacuum, $|0\rangle_1$, for all k and λ. The vacua are related to each other by a Dirac-metric operator in a manner familiar from the scalar case. The metric operator γ is necessary in view of the commutation relations of the Fourier-coefficient operators, eqs. (25) and (26). A one-photon ket is defined by

$$|k\lambda\rangle = a^\dagger_{2k\lambda}|0\rangle_2\,, \tag{32}$$

while the dual one-photon bra is defined by

$$\langle k\lambda| = {}_1\langle 0|a_{1k\lambda}\,, \tag{33}$$

so that

$$\langle k'\lambda'|k\lambda\rangle = \delta_{\lambda\lambda'}\,\delta^3(\vec{k}-\vec{k}')\,, \tag{34}$$

by means of eq. (26) and ${}_1\langle 0|0\rangle_2 = 1$. The metric operator is introduced in order to relate the dual of a ket to its Hermitian conjugate:

$$\langle k\lambda| = (|k\lambda\rangle)^\dagger\gamma\,. \tag{35}$$

Consequently the metric operator is defined to have the properties

$$\gamma|0\rangle_1 = |0\rangle_2\,, \tag{36}$$

$$\gamma^{-1}a_{1k\lambda}\gamma = a_{2k\lambda}\,, \tag{37}$$

$$\gamma^{-1}a_{2k\lambda}\gamma = a_{1k\lambda} \tag{38}$$

with $\gamma = \gamma^\dagger = \gamma^{-1}$. It has a form similar to the metric operator of the scalar case which is given in eq. (36) of ref. (1). The definition of many-photon bras and kets is a direct generalization of eqs. (32), (33) and (35).

The time-ordered propagator which will later be of use in the development of perturbation theory is

$$(39)\qquad G^{22}_{\mu\nu}(x-y) = -i_1\langle 0|T(A_{2\mu}(x)A_{2\nu}(y))|0\rangle_2 = \sin^2\theta_2 G_{F\mu\nu}(x-y) + \cos^2\theta_2 G^*_{F\mu\nu}(x-y),$$

where $G_{F\mu\nu}(x-y)$ is the usual Feynman propagator for photons, while G^*_F is its complex conjugate. Another nonzero free-field propagator, $G^{12}_{\mu\nu}$ is defined as the vacuum expectation value of the time-ordered product of $A_{1\mu}(x)$ and $A_{2\nu}(y)$. It depends on both θ_1 and θ_2. It does not appear in perturbation theory.

Equation (39) shows that $G^{22}_{\mu\nu}$ is a Feynman propagator if $\theta_2 = \pi/2$, and a principal-value propagator if $\theta_2 = \pi/4$. Thus

$$(40)\qquad G^{22}_{\mu\nu}(x-y)|_{\theta_2=\pi/4} = g_{\mu\nu}\int\frac{d^4k}{(2\pi)^4}\exp[-ik\cdot(x-y)]\frac{P}{k^2},$$

where

$$(41)\qquad \frac{P}{k^2} = \frac{1}{2}\left(\frac{1}{k^2+i\varepsilon}+\frac{1}{k^2-i\varepsilon}\right),$$

in the Feynman gauge. This Green's function has previously appeared in action-at-a-distance formulations (3) of classical electrodynamics. The generality of our formulation of quantum field theory allows it to appear as a special case. This does not seem to be possible within the framework of the conventional formulation (7). If we choose $\theta_2 = \pi/4$ and proceed to develop perturbation theory for the full interacting theory then we obtain a quantum-action-at-a-distance theory. We obtain quantum exchange of energy and momentum to which we associate lines in Feynman diagrams. The only difference is in the analytic structure associated with the « epsilonties » of the pole locations. Scattering amplitudes are piecewise analytic in this case (4). More importantly, quanta associated with the electromagnetic field do not appear in asymptotic states, if unitarity is to be maintained. Thus the electromagnetic field does not have its own degrees of freedom, but is constrained to have its source in the charged matter fields which are present. The physical expectations of a quantum action-at-a-distance electrodynamics are therefore satisfied. Except for the absence of « photons » from asymptotic states, and the principal-value nature of the electromagnetic-field propagator, the perturbation theory rules, diagrams and combinatorics are identical to those of QED, although they must be unitarized using the method described in the appendix.

(7) R. Feynman: *Science*, **153**, 699 (1966).

We now turn to the Feynman case ($\theta_2 = \pi/2$). The field $\vec{A}_2$ takes the form

$$\vec{A}_2 = \int \mathrm{d}^3k \sum_{\lambda=1}^{2} \vec{\varepsilon}(k, \lambda)[f_k(x) a_{1k\lambda} + f_k^*(x) a_{2k\lambda}^\dagger] , \tag{42}$$

in the Coulomb gauge. Note that the form of eq. (42) and the commutators, eq. (25) and (26) imply the free-field identity

$$ {}_2\langle 0|A_{2i_1}(x_1) A_{2i_2}(x_2) \dots A_{2i_n}(x_n)|0\rangle_2 = \langle 0|A_{i_1}(x_1) A_{i_2}(x_2) \dots A_{i_n}(x_n)|0\rangle \tag{43}$$

for all spatial indices $i_1, i_2, \dots, i_n$, and all n for which the right-hand side is the vacuum expectation value of n free fields in the usual formulation. This identity is the basis of our claim that the Feynman case of our formulation is *completely* equivalent to QED in its predictions. An important part of establishing the correspondence is the LSZ reduction formulae for photons: for an in photon $(k\lambda)$ we have

$$\begin{aligned} \langle \beta \text{ out}|\alpha(k\lambda) \text{ in}\rangle &= \langle \beta \text{ out}|a_{2k\lambda\,\text{in}}^\dagger|\alpha \text{ in}\rangle = \\ &= \langle \beta - (k\lambda) \text{ out}|\alpha \text{ in}\rangle - \frac{i}{\sqrt{Z_3}} \int \mathrm{d}^4x\, A_{k\lambda}^\mu(x) \overset{\leftarrow}{\Box} \langle \beta \text{ out}|A_{2\mu}(x)|\alpha \text{ in}\rangle \end{aligned} \tag{44}$$

and for an out photon $(k\lambda)$ we have

$$\begin{aligned} \langle \beta(k\lambda) \text{ out}|\alpha \text{ in}\rangle &= \langle \beta \text{ out}|a_{1k\lambda\,\text{out}}|\alpha \text{ in}\rangle = \\ &= \langle \beta \text{ out}|\alpha - (k\lambda) \text{ in}\rangle - \frac{i}{\sqrt{Z_3}} \int \mathrm{d}^4x \langle \beta \text{ out}|A_{2\mu}(x)|\alpha \text{ in}\rangle \overset{\leftarrow}{\Box} A_{k\lambda}^{\mu *}(x) \end{aligned} \tag{45}$$

in the notation of ref. ([5]).

The second quantization of the free-electron field is described in detail in ref. ([1]). We assume an in and out electron field formalism is developed along those lines. Our formulation of spin-one-half fermion theory also allows one to choose Feynman propagators or principal-value propagators for fermions. In order to establish a model equivalent to QED we choose Feynman propagators for the electrons.

The perturbation theory for our model is based on the interaction Lagrangian

$$\mathscr{L}_{\text{int}} = -e_0\, \tilde{\psi}_2 \gamma^0 (\gamma \cdot A_2) \psi_2 . \tag{46}$$

We shall need the LSZ reduction formulae for electrons

$$\begin{aligned} \langle \beta \text{ out}|(ps)\alpha \text{ in}\rangle &= \langle \beta - (ps) \text{ out}|\alpha \text{ in}\rangle - \\ &- \frac{i}{\sqrt{Z_2}} \int \mathrm{d}^4x \langle \beta \text{ out}|\tilde{\psi}_2(x)\gamma^0|\alpha \text{ in}\rangle (-\overleftarrow{i\nabla - m})\, U_{ps}(x) \end{aligned} \tag{47}$$

and

$$(48)\qquad \langle\beta(ps)\,\text{out}|\alpha\,\text{in}\rangle = \langle\beta\,\text{out}|\alpha\,(ps)\,\text{in}\rangle - \\ -\frac{i}{\sqrt{Z_2}}\int d^4x\,\overline{U}_{ps}(x)(i\overrightarrow{\nabla}-m)\langle\beta\,\text{out}|\psi_2(x)|\alpha\,\text{in}\rangle$$

in the notation of ref. ([5]). To remove a positron from an in state, we use

$$(49)\qquad \langle\beta\,\text{out}|(ps)\alpha\,\text{in}\rangle = \langle\beta-(ps)\,\text{out}|\alpha\,\text{in}\rangle + \\ +\frac{i}{\sqrt{Z_2}}\int d^4x\,\overline{V}_{\overline{ps}}(x)(i\overrightarrow{\nabla}-m)\langle\beta\,\text{out}|\psi_2(x)|\alpha\,\text{in}\rangle$$

and, to remove a positron from an out state, we use

$$(50)\qquad \langle\beta(ps)\,\text{out}|\alpha\,\text{in}\rangle = \langle\beta\,\text{out}|\alpha-(ps)\,\text{in}\rangle + \\ +\frac{i}{\sqrt{Z_2}}\int d^4x\langle\beta\,\text{out}|\tilde{\psi}_2\gamma^0|\alpha\,\text{in}\rangle(-i\overleftarrow{\nabla}-m)\,V_{\overline{ps}}(x)\,.$$

In view of the interaction Lagrangian and eqs. (47)-(50) it is clear that only time-ordered products of $\psi_{2\text{in}}$ and $\tilde{\psi}_{2\text{in}}$ appear in the perturbation theory expansion. The Wick theorem, which applies in the present case, reduces all vacuum expectation values of time-ordered products of $\psi_{2\text{in}}$ and $\tilde{\psi}_{2\text{in}}$ to products of the time-ordered two-point function

$$(51)\qquad iS(x-y) = {}_1\langle 0|T\big(\psi_2(x)\,\tilde{\psi}_2(y)\gamma^0\big)|0\rangle_2$$

$$(52)\qquad = i\int\frac{d^4p}{(2\pi)^4}\exp[-ip\cdot(x-y)]\,\frac{p+m}{p^2-m^2+i\varepsilon}\,,$$

where eqs. (51) is in the notation of ref. ([1]).

The above considerations and the familiar development* of the S-matrix as an expansion in vacuum expectation values of time-ordered products of in-field operators lead us to assert that our model electrodynamics is identical to QED in its predictions. In particular

$$(53)\qquad {}_1\langle 0|T\Big(A_{2\mu_1\text{in}}(x_1)\,A_{2\mu_2\text{in}}(x_2)\dots\psi_{2\text{in}}(y_1)\,\psi_{2\text{in}}(y_2)\dots\tilde{\psi}_{2\text{in}}(z_1)\,\tilde{\psi}_{2\text{in}}(z_2)\dots \\ \dots\exp\Big[i\int_{-\infty}^{\infty}dt\,L_{\text{int}}\Big]\Big)|0\rangle_2 = \\ =\langle 0|T\Big(A_{\mu_1\text{in}}(x_1)A_{\mu_2\text{in}}(x_2)\dots\psi_{\text{in}}(y_1)\,\psi_{\text{in}}(y_2)\dots\overline{\psi}_{\text{in}}(z_1)\overline{\psi}_{\text{in}}(z_2)\dots\exp\Big[i\int_{-\infty}^{\infty}dt\,L_{\text{int}}^{\text{QED}}\Big]\Big)|0\rangle\,,$$

where the right-hand side is the corresponding time-ordered product of the conventional formulation of QED with

$$L^{\text{QED}}_{\text{int}} = -i :\!\int \mathrm{d}^3x\, e_0\, \bar{\varphi}_{\text{in}}(\gamma\cdot A_{\text{in}})\varphi_{\text{in}}\!: . \tag{54}$$

Thus, the perturbation theory of our model electrodynamics is term by term equivalent to QED—the equivalent QED term being obtained by ignoring the subscripts 1 and 2 on fields and vacua, and letting $\tilde{\varphi}_2\gamma^0 \to \bar{\varphi}$. Feynman diagrams can be appropriated without modification for use in our formulation. We conclude with a quote from Feynman ([7]): « It always seems odd to me that the fundamental laws of physics can appear in so many different forms that are not apparently identical at first.... Theories of the known which are described by different physical ideas may be equivalent in all their predictions and hence scientifically indistinguishable. However they are not psychologically identical when one is trying to move from that base into the unknown. For different views suggest different modifications ... » In that spirit we turn to the investigation of non-Abelian theories.

3. – Yang-Mills models.

The models we have hitherto investigated are members of a class whose Lagrangians have the form

$$\mathscr{L} = \mathscr{L}_0(\varphi_2) + \mathscr{L}_1(\varphi_1 - \varphi_2)\,, \tag{55}$$

where $\mathscr{L}_0$ has a form which is essentially identical to a Lagrangian of the conventional formulation. For example, in the case of electrodynamics the Lagrangian of eq. (2) can be put in the form of eq. (55), if we let

$$\mathscr{L}_0(A_{2\mu}, \varphi_2) = -\tfrac{1}{4}F^2_{\mu\nu}F^{2\mu\nu} + \tilde{\varphi}_2\gamma^0\big(i(\gamma\cdot\nabla) - e_0(\gamma\cdot A_2) - m\big)\varphi_2 \tag{56}$$

and

$$\begin{aligned}\mathscr{L}_1(A_{1\mu} - A_{2\mu}, \varphi_1 - \varphi_2) &= \\ &= -\tfrac{1}{4}(F^1_{\mu\nu} - F^2_{\mu\nu})^2 + (\tilde{\varphi}_2 - \tilde{\varphi}_1)\gamma^0\big(i(\gamma\cdot\nabla) - m\big)(\varphi_1 - \varphi_2)\,.\end{aligned} \tag{57}$$

At the level of c-number fields it is clear from eqs. (56) and (57) that $A_{2\mu}$ and φ_2 reproduce the usual electrodynamics. At the level of quantum fields one can opt to have a model reproducing the conventional quantum theory—we call it the Feynman case. It maintains the implied decoupling of $A_{1\mu} - A_{2\mu}$ and $\varphi_1 - \varphi_2$ from $A_{2\mu}$ and φ_2. On the other hand, we can take advantage of the larger space of states in the larger-manifestly-indefinite-metric theory and couple the fields together by taking advantage of degrees of freedom in the

Fourier expansion of asymptotic fields and the definition of the vacuum. Since we « perturb around » the asymptotic fields, perturbation theory will reflect this coupling. Our principal-value or quantum action-at-a-distance models are based on this possibility.

We shall use the form of eq. (55) as an ansatz to generate Yang-Mills models in our formulation. This leads to the gauge field Lagrangian

$$\mathscr{L} = -\tfrac{1}{4}\boldsymbol{F}_{2\mu\nu}\cdot\boldsymbol{F}_2^{\mu\nu} + \tfrac{1}{4}(\boldsymbol{G}_{1\mu\nu} - \boldsymbol{G}_{2\mu\nu})\cdot(\boldsymbol{G}_1^{\mu\nu} - \boldsymbol{G}_2^{\mu\nu}), \tag{58}$$

where

$$\boldsymbol{F}_{2\mu\nu} = \partial_\mu \boldsymbol{A}_{2\nu} - \partial_\nu \boldsymbol{A}_{2\mu} + g\boldsymbol{A}_{2\mu}\times\boldsymbol{A}_{2\nu} \tag{59}$$

and

$$\boldsymbol{G}_{i\mu\nu} = \partial_\mu \boldsymbol{A}_{i\nu} - \partial_\nu \boldsymbol{A}_{i\mu} \tag{60}$$

for $i = 1, 2$. Under a local gauge transformation, $S = S(x)$,

$$A_{2\mu} \to A'_{2\mu} = S^{-1}A_{2\mu}S + \frac{i}{g}S^{-1}\partial_\mu S, \tag{61}$$

$$A_{1\mu} \to A'_{1\mu} = A_{1\mu} + A'_{2\mu} - A_{2\mu}, \tag{62}$$

the Lagrangian $\mathscr{L}$ is invariant, where $A_{i\mu} = \boldsymbol{A}_{i\mu}\cdot\boldsymbol{T}$ for $i = 1, 2$ with $\boldsymbol{T}$ a vector composed of matrices representing generators of the gauge group. There is an additional invariance under the local transformation, $\boldsymbol{A}_{1\mu} \to \boldsymbol{A}_{1\mu} + \partial_\mu\boldsymbol{\Lambda}$, which is a straightforward generalization of the Abelian gauge transformation.

The field equations derived from eq. (58), upon varying the action with respect to $\boldsymbol{A}_{1\mu}$ and $\boldsymbol{A}_{2\mu}$, are

$$\partial^\mu(\boldsymbol{G}_{1\mu\nu} - \boldsymbol{G}_{2\mu\nu}) = 0 \tag{63}$$

and

$$(\partial^\mu + g\boldsymbol{A}_2^\mu\times)\boldsymbol{F}_{2\mu\nu} = 0\,. \tag{64}$$

The canonical momentum conjugate to $\boldsymbol{A}_{2\mu}$ is

$$\boldsymbol{\Pi}_{2\mu} = \boldsymbol{F}_{2\mu 0} + \boldsymbol{G}_{1\mu 0} - \boldsymbol{G}_{2\mu 0} \tag{65}$$

and the momentum conjugate to $\boldsymbol{A}_{1\mu}$ is

$$\boldsymbol{\Pi}_{1\mu} = \boldsymbol{G}_{2\mu 0} - \boldsymbol{G}_{1\mu 0}\,. \tag{66}$$

We choose the Coulomb gauge to implement the field quantization. Due to the form of the Lagrangian, we can treat $\boldsymbol{A}_{2\mu}$ and $\boldsymbol{A}_{3\mu} = \boldsymbol{A}_{1\mu} - \boldsymbol{A}_{2\mu}$ as independent fields. Choosing the Coulomb gauge for $A_{2\mu}$, $\vec{\nabla}\cdot\vec{A}_2 = 0$, we can isolate the independent field quantities and establish their equal-time commutation relations in the well-known manner (8). We also can choose to work in the Coulomb gauge of $A_{3\mu}$, $\vec{\nabla}\cdot\vec{A}_3 = 0$ which is equivalent to $\vec{\nabla}\cdot\vec{A}_1 = 0$. This possibility follows from the invariance of $\mathscr{L}$ under the local « Abelian » transformation of $A_{3\mu}$ or $A_{1\mu}$ mentioned above. The resulting equal-time commutation relations are equivalent to

$$[G^{\mathrm{T}}_{\alpha 0 i a}(x),\, A_{\beta j b}(y)] = i(1-\delta_{\alpha 1}\delta_{\beta 1})\,\delta_{ab}\,\Delta^{\mathrm{tr}}_{ij}(x-y)\,, \tag{67}$$

where $\alpha, \beta = 1, 2$; i and j are spatial- and a and b are internal-symmetry indices. G^{T} is the transverse part of G. One can now proceed to linearize the field equations. The Fourier expansions of the solutions of the linearized equations have the form of eqs. (23) and (24), if an internal-symmetry index is appended to the Fourier component operators. The discussion then reduces to the Abelian case considered in the last section. In particular the Green's function of the linearized model which is relevant to perturbation theory is

$$G^{22}_{ab\mu\nu}(x-y) = -i\,{}_1\langle 0|T\big(A_{2\mu a}(x)\,A_{2\nu b}(y)\big)|0\rangle_2\,. \tag{68}$$

It can be expressed as a linear combination of a Feynman propagator and its complex conjugate as in eq. (39). The arbitrary angle can be chosen to give Feynman propagators or principal-value propagators.

The perturbation theory is based on the linearized model. In the Feynman case the LSZ reduction of asymptotic non-Abelian quanta is essentially given in eqs. (44) and (45), if appropriate internal-symmetry indices are introduced. Only fields of « type 2 » are introduced by the LSZ reduction. Furthermore, the cubic and quartic terms of the Lagrangian « interaction » term also involve only fields of type 2. If we restrict all couplings to other fields—such as fermions—to involve only $A_{2\mu}$, then only time-ordered products of $A_{2\mu}$ appear in the perturbation theory expansion of the S-matrix. The consequence of this development is that one has constructed a model which makes predictions which are identical to a conventional formulation based on a Lagrangian of the usual type

$$\mathscr{L} = -\tfrac{1}{4}\boldsymbol{F}_{\mu\nu}\cdot\boldsymbol{F}^{\mu\nu}\,. \tag{69}$$

This can be verified in a path integral approach based on the Lagrangian of eq. (58).

The point of our formulation is that one can make other choices—such as the principal-value propagator choice. (Note that the path integral formulation

(8) E. ABERS and B. W. LEE: *Phys. Rep.*, **9** C, 1 (1973).

only has a formal validity in these cases, since the Wick rotation to Euclidean co-ordinates is highly nontrivial.) The choice of principal-value propagators has several important consequences.

In the appendix we show how to calculate the unitary physical S-matrix in this case. We show that loops of principal-value propagators do not contribute to the absorptive parts of S-matrix elements. As a result Faddeev-Popov ghost loops are not needed to maintain unitarity (in any gauge). We also point out a radical possibility: a unitary S-matrix can be defined in which all non-Abelian boson loops are excluded. Thus the non-Abelian boson sector consists solely of tree diagrams. This is a quantum analogue of the absence of self-interactions in classical action-at-a-distance theories.

4. – Non-Abelain model of the strong interactions.

Several years ago a non-Abelian model of the strong interaction was constructed (9) which had manifest quark confinement and a linear potential. The model was based on higher-order field equations. Such equations lead to well-known unitarity problems, if the usual quantization program is implemented. The author chose to avoid the unitarity problems through an *ad hoc* quantization procedure which gave principal-value propagators. The result was a model having unconventional analyticity properties. Despite the successful confinement scheme in this model, there was some reason for uneasiness, due to the unusual form of the Lagrangian (where two sets of fields were associated with the gluons) and the *ad hoc* method of quantization.

The new framework we have developed in ref. (1) and this paper eliminates these sources of concern. Our formulation is based on fundamental ground—the need for an acceptable physical particle interpretation of quantum field theory in flat and curved space-time. We shall see that the two-field Lagrangian of the strong-interaction model lies within the general framework we have established. The principal-value propagators, required to have a unitary model, will also be obtained in a straightforward manner.

Since the strong-interaction model is described in some detail in ref. (9), we shall only outline the aspects necessary to connect that model with the present work.

The strong-interaction Lagrangian departs slightly from the ansatz of eq. (55) (mainly in the replacement $\boldsymbol{G}^1_{\mu\nu}\cdot\boldsymbol{G}^{1\mu\nu}\to\Lambda^2\boldsymbol{A}_{1\mu}\cdot\boldsymbol{A}^\mu_1$):

$$\mathscr{L} = -\tfrac{1}{2}\boldsymbol{F}^1_{\mu\nu}\cdot\boldsymbol{F}^{2\mu\nu} - \tfrac{1}{2}\Lambda^2\boldsymbol{A}_{1\mu}\cdot\boldsymbol{A}^\mu_1 + \bar{\psi}_2\gamma^0\big(i(\gamma\cdot\nabla) + g(\gamma\cdot\boldsymbol{A}_2) - m\big)\psi_2 - (\bar{\psi}_1 - \bar{\psi}_2)\gamma^0\big(i(\gamma\cdot\nabla) - m\big)(\psi_1 - \psi_2)\,. \tag{70}$$

(9) S. BLAHA: *Phys. Rev. D*, **10**, 4268 (1974); **11**, 2921 (1975). It should be noted that the claim that loops of principal-value propagators are identically zero is not true. In the case of one-loop diagrams only the real part is zero.

where ψ_1 and ψ_2 are the quark fields and

$$F^2_{\mu\nu} = \partial_\mu A_{2\nu} - \partial_\nu A_{2\mu} + g A_{2\mu} \times A_{2\nu} \tag{71}$$

and

$$F^1_{\mu\nu} = D_\mu A_{1\nu} - D_\nu A_{1\mu} \tag{72}$$

with the covariant derivative defined by

$$D_\mu = \partial_\mu + g A_{2\mu} \times . \tag{73}$$

The Lagrangian is invariant under the local color SU_3 gauge transformation

$$\psi_2 \to S^{-1} \psi_2 , \tag{74}$$

$$\psi_1 \to \psi_1 + (S^{-1} - I) \psi_2 , \tag{75}$$

$$A_{1\mu} \to S^{-1} A_{1\mu} S , \tag{76}$$

$$A_{2\mu} \to S^{-1} A_{2\mu} S + \frac{i}{g} S^{-1} \partial_\mu S , \tag{77}$$

with $A_{i\mu} = \boldsymbol{A}_{i\mu} \cdot \boldsymbol{T}$, where $\boldsymbol{T}$ is a vector composed of matrices representing SU_3 generators. The equations of motion and the Coulomb gauge quantization are discussed in detail in ref. [9]. (The subscripts, 1 and 2, on the gluon fields are reversed in ref. [9] relative to the present development.)

The essential feature of the model can be seen in the linearized field equations

$$\Box \boldsymbol{A}_{2\mu} = -\Lambda^2 \boldsymbol{A}_{1\mu} , \tag{78}$$

$$\Box \boldsymbol{A}_{1\mu} = -g \boldsymbol{J}_\mu , \tag{79}$$

where $\boldsymbol{J}_\nu$ is the color quark current. Equations (78) and (79) show an apparent mass term, $\Lambda^2(\boldsymbol{A}_{1\mu})^2$, actually leads to a higher-order derivative field equation. The implications of this feature for quark confinement are discussed in ref. [9]. The discussion is based on the assumption that all gluon propagators are principal-value propagators. We shall now show that the propagators for color gluons in the linearized model can be principal-value propagators.

We work in the Coulomb gauge, $\vec{\nabla} \cdot \vec{\boldsymbol{A}}_2 = 0$. A mode expansion of the gluon fields must be consistent with the free linearized field equations and the equal-time commutation relations,

$$[F^{\alpha T}_{0ia}(x), A_{\beta jb}(y)] = i\delta_{ab}(1 - \delta_{\alpha\beta})\delta^{tr}_{ij}(\vec{x} - \vec{y}) \tag{80}$$

(cf. eqs. (65) and (66) of ref. [9]). These conditions and the requirement of a unitary theory imply the Fourier expansions

$$(81)\qquad \vec{A}_{1a}(x) = \int \frac{\mathrm{d}^3k}{\sqrt{2}} \sum_{\lambda=1}^{2} \vec{\varepsilon}(k,\lambda)[(a_{1k\lambda a} + a_{2k\lambda a}) f_k(x) + (a^\dagger_{1k\lambda a} + a^\dagger_{2k\lambda a}) f^*_k(x)]\,,$$

$$(82)\qquad \vec{A}_{2a}(x) = \int \frac{\mathrm{d}^3k}{\sqrt{2}} \sum_{\lambda=1}^{2} \vec{\varepsilon}(k,\lambda)[(a_{2k\lambda a} - a_{1k\lambda a}) f_k(x) + (a^\dagger_{2k\lambda a} - a^\dagger_{1k\lambda a}) f^*_k(x)] +$$

$$+ \Lambda^2 \int \frac{\mathrm{d}^4k}{\sqrt{2}} 2\omega_k \theta(k_0) \delta'(k^2) \sum_{\lambda=1}^{2} \vec{\varepsilon}(k,\lambda)[(a_{1k\lambda a} + a_{2k\lambda a}) f_k(x) + (a^\dagger_{1k\lambda a} - a^\dagger_{2k\lambda a}) f^*_k(x)]\,,$$

where $\delta'(k^2) = \partial\delta(k^2)/\partial k^2$ and $\omega_k = |\vec{k}|$. From eqs. (81) and (82) we can determine the Green's functions of the linearized model:

$$(83)\qquad iG^{12}_{\mu\nu ab}(x-y) = \langle 0|T(A_{1\mu a}(x)A_{2\nu b}(y))|0\rangle =$$

$$= -i\delta_{ab}\int \frac{\mathrm{d}^4k}{(2\pi)^4} \exp[-ik\cdot(x-y)]\, r^{12}_{\mu\nu} P\frac{1}{k^2}$$

and

$$(84)\qquad iG^{22}_{\mu\nu ab}(x-y) = \langle 0|T(A_{2\mu a}(x)A_{2\nu b}(y))|0\rangle =$$

$$= i\Lambda^2\delta_{ab}\int \frac{\mathrm{d}^4k}{(2\pi)^4} \exp[-ik\cdot(x-y)]\, r^{22}_{\mu\nu} P\frac{1}{k^4}\,,$$

where

$$(85)\qquad P\frac{1}{k^{2N}} = \frac{1}{2}\left[\frac{1}{(k^2+i\varepsilon)^N} + \frac{1}{(k^2-i\varepsilon)^N}\right]$$

and where $r^{12}_{\mu\nu}$ and $r^{22}_{\mu\nu}$ are gauge-dependent tensors. While the path integral formulation is only of formal value for the present model, it can be used to determine the form of the Green's functions for any choice of gauge. It can also be used to generate the perturbation theory rules for the model. The character of the perturbation theory is described in ref. (9). Confinement arises through the Schwinger mechanism in a manner reminiscent of the two-dimensional Schwinger model. The mass scale is set by Λ. The Lagrangian of eq. (70) leads to an interaction between quarks which has the linear potential r as its Coulomb potential in lowest order. The r potential has had notable success in the explication of the charmonium system. A r^{-1} term may also be needed to successfully describe charmonium. Such a term can be introduced in our model by adding another interaction term

$$(86)\qquad \mathscr{L}_r = g'\,\bar{\psi}_2\gamma^a(\gamma\cdot A_1)\psi_2$$

to eq. (70). Since $A_{1\mu}$ transforms homogeneously under a gauge transformation, the gauge invariance of the Lagrangian is not altered by the addition of this term.

In conclusion, we have established the basis for a manifestly quark-confining model of the strong interaction within the framework of our formulation of quantum field theory. This example illustrates the value of our formulation in the construction of unitary quantum field theories with higher-order derivative field equations. Another interesting application of this approach would be to quantum gravity which may have its renormalization problems ameliorated by introducing higher-order derivatives. Certain formal similarities, and this possibility, have led the author to propose a unified model of vierbein gravity and the strong interaction [10]. The higher-order derivative gravity terms which lead to power counting renormalizability for the quantum gravity sector are linked to the higher-order derivative strong-interaction sector terms which lead to quark confinement. The recent discovery of important spin effects in high-energy p-p scattering [11] is suggestive in view of the presence of strong spin-spin interactions in unified models of the strong-interaction and vierbein gravity.

5. – Model of quantum gravity.

In this section we develop a trivially renormalizable model of quantum gravity (based on the Einstein Lagrangian) which is coupled to an external classical source. In addition to obtaining a physically interesting model, we shall see the utility of principal-value propagators in ameliorating divergence problems. The self-interactions of the « gravitons » will be reduced if principal-value propagators are used. This corresponds to the absence of self-interactions in classical action-at-a-distance models.

The model which we shall develop lies within the framework established above. It can be described as quantum action-at-a-distance gravity. We begin with the Einstein action

$$I = -2\varkappa^{-2}\int d^4x\, g^{\frac{1}{2}} R \tag{87}$$

with $\varkappa = 32\pi G$, where G is Newton's constant, $g = \det g_{\mu\nu}$, $R = g^{\mu\nu} R^{\lambda}_{\mu\nu\lambda}$ and where

$$R^{\alpha}_{\mu\nu\lambda} = \partial_\nu \Gamma^{\alpha}_{\mu\lambda} + \Gamma^{\alpha}_{\nu\beta}\Gamma^{\beta}_{\mu\lambda} - (\nu \Leftrightarrow \lambda) \tag{88}$$

and

$$\Gamma^{\alpha}_{\mu\nu} = \frac{1}{2} g^{\alpha\beta}[\partial_\mu g_{\nu\beta} + \partial_\nu g_{\mu\beta} - \partial_\beta g_{\mu\nu}]\,. \tag{89}$$

---

[10] S. Blaha: *Lett. Nuovo Cimento*, **18**, 60 (1977).

[11] J. R. O'Fallon, L. G. Ratner, P. F. Schultz, K. Abe, R. C. Fernow, A. D. Krisch, T. A. Mulera, A. J. Salthouse, B. Sandler, K. M. Terwilliger, D. G. Crabb and P. H. Hansen: *Phys. Rev. Lett.*, **39**, 733 (1977).

Our approach is analogous to Feynman's covariant quantization around a flat background field ([12]). We therefore let

$$g_{\mu\nu} = \eta_{\mu\nu} + \varkappa h^2_{\mu\nu}, \tag{90}$$

where $\eta_{\mu\nu} = (-1, 1, 1, 1)$. We then introduce a second tensor field, $h^1_{\mu\nu}$; and choose the action for our model to be (cf. eq. (55))

$$I = \int d^4x [\tfrac{1}{2}(h^3_{\mu\nu,\sigma})^2 - (h^3_\mu)^2 - \tfrac{1}{2}(h^3_{,\mu})^2 - 2\varkappa^{-2} g^{\frac{1}{2}} R], \tag{91}$$

where $h^3_{\mu\nu} = h^1_{\mu\nu} - h^2_{\mu\nu}$, $h^3_\mu = h^3_{\mu\nu,\nu}$, $h^3_{,\mu} = h_{\nu\nu,\mu} = \partial_\mu h$ and where the last term on the right-hand side of eq. (91) is expanded in $h^2_{\mu\nu}$ by using eq. (90). The gauge invariance of the Lagrangian is maintained by requiring $h^1_{\mu\nu}$ and $h^2_{\mu\nu}$ to satisfy the same transformation law

$$h^i_{\mu\nu} \to h^{i\prime}_{\mu\nu} = h^i_{\mu\nu} - \partial_\nu \varepsilon_\mu - \partial_\mu \varepsilon_\nu, \tag{92}$$

where ε_μ are four small, but otherwise arbitrary, functions of the co-ordinates. We introduce a de Donder harmonic gauge fixing term

$$\mathscr{L}_D = -\frac{1}{2\gamma}\left(h^2_\mu - \frac{1}{2} h^2_{,\mu}\right)^2 + \frac{1}{2\gamma}\left(h^3_\mu - \frac{1}{2} h^3_{,\mu}\right)^2, \tag{93}$$

in order to obtain a regular kinetic matrix from the quadratic terms of the augmented Lagrangian. The quadratic terms are (if we let $\gamma = \frac{1}{2}$)

$$\mathscr{L}_{\text{quad}} = -h^1_{\alpha\beta,\mu} h^2_{\alpha\beta,\mu} + \tfrac{1}{2} h^1_{,\mu} h^2_{,\mu} + \tfrac{1}{2}(h^1_{\alpha\beta,\mu})^2 - \tfrac{1}{4}(h^1_{,\mu})^2 . \tag{94}$$

These terms plus the higher-order «interaction» terms can be introduced into a suitable path integral ([13]) expression and the perturbation theory rules can be developed. The path integral approach is only of formal significance, in general, in our models. It gives the correct algebraic expressions and combinational rules. The nontriviality of Wick rotation to Euclidean co-ordinates in any case, but the Feynman case, precludes attributing anything more than formal value to the path integral approach. Of course, the applicability of the path integral formalism to the Feynman case, and the absence of any difference at the algebraic level (before integrations) between the Feynman and principal-value cases, means that the path integral formalism also describes the combinatorics of the principal-value case.

([12]) R. Feynman: *Acta Phys. Polonica*, **24**, 697 (1963).
([13]) V. N. Popov and L. Fadeev: Kiev Report. No. ITP 67-36 (1967).

In order to have a nontrivial model in the principal-value case, we shall assume the gravitational field is coupled to an external classical source. Note that covariance requires that only $h^2_{\mu\nu}$ couples to that source. This fact, plus the absence of $h^1_{\mu\nu}$ in higher-order terms, implies that the only free-field propagator necessary to develop perturbation theory is the vacuum expectation value of the time-ordered product of free $h^2_{\mu\nu}$ fields:

$$(95)\qquad iG^{22}_{\mu\nu,\varrho\sigma} = {}_1\langle 0|T\big(h^2_{\mu\nu}(x)h^2_{\varrho\sigma}(y)\big)|0\rangle_2 = \\ = -\frac{i}{2}(\eta_{\mu\varrho}\eta_{\nu\sigma}+\eta_{\mu\sigma}\eta_{\nu\varrho}-\eta_{\mu\nu}\eta_{\varrho\sigma})\int\frac{d^4k}{(2\pi)^4}\,\frac{\exp[-ik\cdot(x-y)]}{k^2}\,.$$

If one chooses the propagator in eq. (95) to be a Feynman propagator, then the usual model of quantum gravity results with its attendant renormalizability problems.

We shall consider the case, in which G^{22} is a principal-value propagator. One can establish this case by following a canonical quantization procedure which is completely analogous to those of models considered earlier. The propagation of « gravitons » by principal-value propagators has important consequences for perturbation theory. All graviton loops can be excluded from the physical S-matrix without leading to unitarity problems (cf. appendix). As a result, only tree diagrams occur and unitarity requires that all gravitons are absorbed—either within trees or on the external classical source. Thus we have a model quantum gravity with no renormalization problem. In the classical limit the model becomes Einstein's classical theory of gravity ([14]). There is, of course, the question of whether the classical limit is a retarded model of gravity or not. The answer lies in cosmology—is the absorber mechanism operative?

An action-at-a distance model of gravity with absorber seems to be very much in the spirit of Mach's principle. The usual approach—where the gravitational field is treated as having its own degrees of freedom—almost invariably leads to the surrepitious introduction of absolute space. The reason is simple. The solution of the field equations for a localized mass distribution is asymptotically flat. The only apparent way to avoid this problem is to say Mach restricts us to the class of closed-universe solutions.

In an action-at-a-distance gravity ([15]) the metric field exists only in the presence of matter, so that any asymptotic flatness of solutions is a mathematical

([14]) There is an opinion which is sometimes expressed that a tree diagram model is equivalent to a classical theory. This is not true. The correct view is that the classical limit of a tree diagram model is the corresponding classical theory. Cf. S. DESER's paper in *Quantum Gravity*, edited by C. J. ISHAM, R. PENROSE and D. W. SCIAMA (Oxford, 1975).

([15]) A. WHITEHEAD: *The Principle of Relativity* (Cambridge, 1922).

artifact. The metric properties of space are a consequence of the presence of mass-energy. (Consistent with this view, we note that mass-energy must be present to measure the metric properties of space.)

The ultimate realization of Mach's principle from the present viewpoint requires that inertia be the result of the gravitational effects of distant matter. Preliminary work ([16]) in this direction can be easily incorporated within the framework of an action-at-a-distance model of gravity. In fact, Sciama's model of the inertial effects of the distant stars displays a close analogy to the Feynman-Wheeler absorber model.

In closing we remark that the introduction of quantum matter fields in our action-at-a-distance quantum gravity appears to result in a nonrenormalizable model. It seems that a higher-order derivative Lagrangian of the type mentioned in the last section may be necessary for a fully renormalizable quantum gravity. Of course, that would also be a quantum action-at-a-distance model.

APPENDIX

In this appendix we define a unitary physical S-matrix for action-at-a-distance quantum field theories. The usual formal definition of the S-matrix is unacceptable because it introduces negative-metric states which lead to negative probabilities.

We begin by defining the set of physical asymptotic states to be those states of positive metric which do not contain quanta of action-at-a-distance fields. In an action-at-a-distance electrodynamics the set of physical states includes all electron and positron states not containing «action at a distance» photons. The problem is to define a unitary S-matrix taking «in» physical states to «out» physical states.

We choose to use a variation of Bogoliubov's procedure ([17]) for defining a unitary physical S-matrix as implemented by SUDARSHAN and co-workers ([18]). The starting point is the expansion of the S-matrix in «old fashioned» perturturbation theory:

$$S_{\beta\alpha} = \delta_{\beta\alpha} - i(2\pi)^4\delta^4(P_\beta - P_\alpha)T_{\beta\alpha} \tag{A.1}$$

with

$$T_{\beta\alpha} = \langle\beta|H_I|\alpha\rangle + \sum_N \frac{\langle\beta|H_I|N\rangle\langle N|H_I|\alpha\rangle}{E_\alpha - E_N + i\varepsilon} + \dots, \tag{A.2}$$

([16]) D. W. SCIAMA: *Roy. Astron. Soc. Month. Not.*, **113**, 34 (1953).

([17]) N. N. BOGOLIUBOV: *Annales of Invitational Conference on High-Energy Physics*, CERN (1958).

([18]) E. C. G. SUDARSHAN: *Fields and Quanta*, **2**, 175 (1972); C. A. NELSON: Louisiana State University preprint (1972); J. L. RICHARD: *Phys. Rev. D*, **7**, 3617 (1973); C. C. CHIANG: University Göteborg preprint (1972), and references therein.

where H_I is the interaction Hamiltonian. We wish to modify eq. (A.2) so that states with «action-at-a-distance quanta» do not contribute to the absorptive part of amplitudes. This would allow unitarity to be maintained under the restriction of unitarity sums to the set of physical states. SUDARSHAN and co-workers ([18]) have shown that this can be done by taking the energy denominator factor of unphysical states in principal value:

$$(A.3) \qquad T_{\beta\alpha} = \langle\beta|H_I|\alpha\rangle + \sum_p \frac{\langle\beta|H_I|p\rangle\langle p|H_I|\alpha\rangle}{E_\alpha - E_p + i\varepsilon} + \sum_u \langle\beta|H_I|u\rangle\langle u|H_I|\alpha\rangle P\frac{1}{E_\alpha - E_u} + \dots,$$

where p labels physical states and u labels unphysical states (*i.e.* those containing action-at-a-distance quanta in our case). Equation (A.3) defines the physical T-matrix in our models. It could serve as the basis for calculating S-matrix elements.

However, we would like to re-express this S-matrix in terms of covariant perturbation theory. In order to do this we shall take advantage of Weinberg's study ([19]) of the infinite-momentum frame limit of old-fashioned perturbation theory, and Chang and Ma's realization of infinite-momentum frame results by a change of variable ([20]).

Since the difference between the physical T-matrix and the unmodified T-matrix lies only in the character of the singularity in the energy denominator, the algebraic development of Weinberg, and his power counting arguments, in particular, can be taken over without change to our case. (We also will ignore possible complications due to interchanging the order of integration and the $P\to\infty$ limit. Actually we define our T-matrix to be the result of this procedure.) The result is that Weinberg's rules apply in the present case also —except that Weinberg's rule (*d*) must be modified, so that an energy denominator is taken in principal value, if the corresponding intermediate state is unphysical—*i.e.* contains action-at-a-distance quanta.

CHANG and MA showed that it was not necessary to go to the infinite-momentum frame limit in order to realize Weinberg's rules. They showed that the expression of the usual Feynman rules in terms of infinite momentum frame variables, $\eta = p^0 + p^3$ and $s = p^0 - p^3$ for each momentum p^μ, led to the same results as Weinberg's rules. This fact can be used to advantage in the case of action-at-a-distance models. S-matrix elements can be calculated in our models from the usual Feynman rules in the following way: i) for a given diagram follow the usual Feynman rules to obtain a S-matrix contribution using infinite-momentum frame variables *à la* Chang and Ma—in particular, associate Feynman propagators with action-at-a-distance particle lines; ii) evaluate the «energy» integrals by complex integration; iii) the result will be a series of terms corresponding to different intermediate states in old-fashioned perturbation theory language. Those denominators corresponding to intermediate states with principal-value quanta should be taken in principal value. Those denominators corresponding to physical intermediate states should remain unmodified.

([19]) S. WEINBERG: *Phys. Rev.*, **150**, 1313 (1966).
([20]) S. J. CHANG and S. K. MA: *Phys. Rev.*, **180**, 1506 (1969).

Let us examine the results of this procedure for some simple cases. First, consider the pole diagram for two-particle–to–two-particle scattering. If the pole corresponds to a principal-value particle the amplitude will be proportional to

$$P\frac{1}{q^2-m^2}, \tag{A.4}$$

using the above rules ([21]). This result also follows from explicit evaluation of the diagram in canonical perturbation theory, using eq. (1) of the text for $\theta=\pi/4$.

Next consider a second-order self-energy correction due to the emission and absorption of a principal-value quantum by a particle (propagating via Feynman propagators). Our modified Feynman rules require us to evaluate an integral

$$I=i\int d^4k\,\frac{1}{k^2-m^2+i\varepsilon}\,\frac{1}{(p-k)^2-\mu^2+i\varepsilon}, \tag{A.5}$$

using-infinite-momentum frame variables:

$$\eta'=k^0+k^3 \qquad s'=k^0-k^3 \qquad \vec{q}=(k^1,k^2) \tag{A.6}$$

and $p=(s,\eta,\vec{0})$. The result is

$$I=\frac{\pi}{2}\int d^2q\int_0^\eta \frac{d\eta'}{\eta'(\eta-\eta')}\,\frac{1}{s-(\vec{q}^{\,2}+\mu^2)/(\eta-\eta')-(\vec{q}^{\,2}+m^2)/\eta'+i\varepsilon} \tag{A.7}$$

which must be modified to

$$I'=\frac{\pi}{2}\int d^2q\int_0^\eta \frac{d\eta'}{\eta'(\eta-\eta')}\,P\,\frac{1}{s-(q^2+\mu^2)/(\eta-\eta')-(q^2+m^2)/\eta'}. \tag{A.8}$$

In this case we see that $2I'=I+I^*$. The propagation of the particle which would ordinarily propagate via Feynman propagators is manifestly different in states where virtual action-at-a-distance quanta are present. This is necessary in order to define a unitary physical S-matrix. Thus the physical S-matrix only agrees with the canonically defined S-matrix for one–principal-value particle intermediate states and states without principal value particles.

If we consider a non-Abelian action-at-a-distance gauge theory coupled to an external classical source, it is possible to define the physical S-matrix in

---

([21]) Compare to eqs. (5) and (6) of ref. ([9]).

a different manner from the above. Consider the usual definition of the S-matrix and in particular the set of tree diagrams with no external principal-value particle lines. It is easy to verify that the absorptive part of any tree diagram is zero *in the physical region* since

$$\text{Abs}\left[P\,\frac{1}{k^2-m^2}\right]=0\,. \tag{A.9}$$

As a result the set of tree diagrams defines a unitary (gauge invariant) S-matrix. (This should be contrasted with the usual theory where taking the absorptive part of a tree diagram introduces states containing non-Abelian bosons which in turn introduce loops and Fadeev-Popov ghosts.) In the present case no Fadeev-Popov ghost loops are needed to maintain unitarity ([22]). Thus we wind up with a loopless, tree diagram model. The possibility of limiting the self-interaction of fields in this way allows us to develop a renormalizable model of quantum gravity.

Finally we note that the lack of contributions to unitarity sums from intermediate states containing principal-value particles allows us to avoid the introduction of Fadeev-Popov ghosts in *all* non-Abelian action-at-a-distance models.

([22]) Cf. ref. ([12]) and B. W. LEE and J. ZINN-JUSTIN: *Phys. Rev. D*, **5**, 3121 (1972).

● RIASSUNTO (*)

Si formulano teorie di gauge nel sistema di una generalizzazione della teoria quantistica dei campi. In particolare si discutono modelli di teorie di elettrodinamica e di Yang-Mills, un modello dell'interazione forte con derivate di ordine più alto e confinamento dei quark, e un modello rinormalizzabile di gravità quantistica pura con una Lagrangiana di Einstein. Nel caso dell'elettrodinamica si mostra che due modelli sono possibili: uno con predizioni che sono identiche a QED e uno che è un modello quantistico di azione a distanza dell'elettrodinamica. Nel caso delle teorie di Yang-Mills si può costruire un modello che è identico per quanto riguarda le predizioni a qualsiasi modello convenzionale o modello di azione a distanza. Nel secondo caso è possibile eliminare tutti i cappi di particelle di Yang-Mills (in tutti i gauge) in una maniera consistente con l'unitarietà. Esiste una variazione dei modelli di Yang-Mills nella nostra formulazione che ha equazioni di campo con derivate ad ordine più alto. È unitario ed ha probabilità positive. Può essere usato per costruire un modello d'interazioni forti che ha un potenziale lineare e manifesta confinamento dei quark. Infine si mostra come costruire un modello di azione a distanza della gravità quantistica pura (il cui limite classico è la dinamica della Lagrangiana di Einstein) accoppiato ad una sorgente esterna classica. Il modello è grossolanamente rinormalizzabile.

(*) *Traduzione a cura della Redazione.*

Новый подход к калибровочным теориям поля.

Резюме (*). — Мы формулируем калибовочные теории в рамках обобщения квантовой теории поля. В частности, мы обсуждаем модели электродинамики и теорий Янга-Миллса, модель сильных взаимодействий с производными высших порядков и удержанием кварков и перенормируемую модель для чистой квантовой гравитации с Лагранжианом Эйнштейна. Мы показываем, что в случае электродинамики возможны две модели: одна модель имеет предсказания, которые идентичны предсказаниям квантовой электродинамики, и другая модель, которая представляет модель элзктродинамики с квантовым действием на расстоянии. В случае теорий Янга-Миллса мы можем сконструировать модель, которая идентична по предсказаниям любой общепринятой модели или модели с квантовым действием на расстоянии. Во втором случае имеется возможность исключить все петли для частиц Янга-Миллса (во всех калибровках). Наша формулировка содержит изменение моделей Янга-Миллса, которое включает уравнения поля с производными высших порядков. Наша модель является унитарной и имеет положительные вероятности. Наша формулировка может быть использована для конструирования модели сильных взаимодействий, которая имеет линейный потенциал и обеспечивает удержание кварков. Мы показываем, как сконструировать модель действия на расстоянии для квантовой гравитации, связанной с внешним классическим источником. Предложенная модель является перенормируемой.

(*) *Переведено редакцией.*

Direttore responsabile: CARLO CASTAGNOLI

Stampato in Bologna dalla Tipografia Compositori coi tipi della Tipografia Monograf
Questo fascicolo è stato licenziato dai torchi il 17-I-1979

Appendix C. PseudoQuantization Paper

This refereed paper is S. Blaha, Phys. Rev. **D17**, 994 (1978). Reprinted with the kind permission of Physical Review D.

PHYSICAL REVIEW D VOLUME 17, NUMBER 4 15 FEBRUARY 1978

Embedding classical fields in quantum field theories

Stephen Blaha*
Physics Department, Syracuse University, Syracuse, New York 13210
(Received 2 August 1976; revised manuscript received 7 November 1977)

We describe a procedure for quantizing a classical field theory which is the field-theoretic analog of Sudarshan's method for embedding a classical-mechanical system in a quantum-mechanical system. The essence of the difference between our quantization procedure and Fock-space quantization lies in the choice of vacuum states. The key to our choice of vacuum is the procedure we outline for constructing Lagrangians which have gradient terms linear in the field variables from classical Lagrangians which have gradient terms which are quadratic in field variables. We apply this procedure to model electrodynamic field theories, Yang-Mills theories, and a vierbein model of gravity. In the case of electrodynamics models we find a formalism with a close similarity to the coherent-soft-photon-state formalism of QED. In addition, photons propagate to $t = +\infty$ via retarded propagators. We also show how to construct a quantum field for action-at-a-distance electrodynamics. In the Yang-Mills case we show that a previously suggested model for quark confinement necessarily has gluons with principal-value propagation which allows the model to be unitary despite the presence of higher-order-derivative field equations. In the vierbein-gravity model we show that our quantization procedure allows us to treat the classical and quantum parts of the metric field in a unified manner. We find a new perturbation scheme for quantum gravity as a result.

I. INTRODUCTION

The relation between classical and quantum systems has been a subject of continuing interest over the years: First, in the original development of quantum mechanics, second, in the study of the classical limit and infrared divergences of quantum-electrodynamic processes,[1,2] and third, in recent attempts to construct strong-interaction models of quark confinement which are for the most part either classical field theory models in search of quantization[3] or quantized gluon models wherein quark confinement is a consequence of infrared behavior.[4,5]

We will describe a new quantization procedure (called pseudoquantization) for field theory which is the analog of Sudarshan's method for embedding a classical-mechanical system in a quantum-mechanical system. It can be used with advantage to either embed a classical field theory in a quantum field theory in such a way as to maintain the classical character of the embedded fields (while studying the interaction between the classical and quantum sectors on essentially the same footing), or to quantize a class of field theories, members of which have been used as models for gravity and as models for the strong interaction with quark confinement.[7-9]

We shall begin (Sec. II) by pseudoquantizing a classical simple harmonic oscillator. This case is of particular importance because of the analogy between the mode amplitudes of a quantum field and the coordinates of a set of simple harmonic oscillators which we will take advantage of in later sections.

In Sec. III we describe the pseudoquantization procedure for field theory. We apply it to electrodynamic models and show that the propagation of photons to $t=+\infty$ is necessarily retarded in this formalism. Further, we display a close analogy between the present formalism and the coherent-soft-photon-state formalism[10] of QED.

In Sec. IV we apply the pseudoquantization procedure to a classical Yang-Mills field. The resulting field theory (with a slight but important modification) has been used as a model for the strong interactions with quark confinement.[7-9] We also apply the pseudoquantization procedure to a vierbein model of gravity and obtain a new perturbation theory for quantum gravity.

In Sec. V we show that principal-value propagators naturally arise in certains sectors of pseudoquantized theories thus verifying an *ad hoc* procedure devised to unitarize a model of quark confinement.[7-9] We also show how to construct a quantum version of action-at-a-distance electrodynamics.

We shall now briefly outline the procedure for embedding a classical-mechanical system in a quantum system.[6] Consider a classical Hamiltonian system with one degree of freedom, and commuting canonical variables, x_1 and p_1, which have the equations of motion

$$\dot{x}_1 = -i[x_1, \hat{H}], \tag{1}$$

$$\dot{p}_1 = -i[p_1, \hat{H}], \tag{2}$$

where defining

$$\hat{H} = -i\left(\frac{\partial H(x_1,p_1)}{\partial p_1}\frac{\partial}{\partial x_1} - \frac{\partial H(x_1,p_1)}{\partial x_1}\frac{\partial}{\partial p_1}\right) \tag{3}$$

allows us to write Hamilton's equations in commutator form. With Sudarshan[6] we define

$$x_2 = i\frac{\partial}{\partial p_1} \tag{4}$$

and

$$p_2 = -i\frac{\partial}{\partial x_1} \tag{5}$$

so that

$$[x_1, x_2] = [p_1, p_2] = 0\,, \tag{6}$$

$$[x_1, p_2] = [x_2, p_1] = i\,, \tag{7}$$

and $\hat{H}$ can now be taken to be the operator

$$\hat{H} = \frac{\partial H(x_1, p_1)}{\partial p_1} p_2 + \frac{\partial H(x_1, p_1)}{\partial x_1} x_2\,. \tag{8}$$

It is now apparent that we can take the above quantities and equations of motion to describe a quantum mechanical system with two degrees of freedom in the "coordinate" representation where the "coordinates" are (x_1, p_1) and the canonical momenta are $\Pi = (p_2, -x_2)$. As we will see below the linearity of $\hat{H}$ in the momenta is crucial for the maintenance of the classical character of x_1 and p_1, and for the observability of the phase-space trajectory. Since we choose to identify the physical observables with the commutative algebra of the coordinate operators, x_1 and p_1, we are led to impose the superselection condition that the momenta, Π, are unobservable. As a result the Hamiltonian and other generators of canonical transformations, which are all linear in the momenta, are also unobservable. However, in each case there is an associated dynamical quantity which is observable.

The required unobservability of the momenta restricts the form of the interaction between a classical-made-quantum system and an inherently quantum system to

$$H_{\text{int}} = \Phi_1 x_2 + \Phi_2 p_2 + X\,, \tag{9}$$

where Φ_1, Φ_2, and X are functions of x_1, p_1, and the quantum system variables. The commutation relations of these functions are also constrained[6] by the superselection rule and the commutativity of the classical variables, x_1 and p_1, and their time derivatives. In the next section we will study the simple harmonic oscillator in order to exemplify the quantum-mechanical case described above and also for direct use in the field-theoretic generalizations of subsequent sections.

II. SIMPLE HARMONIC OSCILLATOR

In this section we discuss the embedding of a classical simple harmonic oscillator in a quantum system. We shall see that the space of states for the indefinite-metric classical-made-quantum system is far larger than the set of states of a classical harmonic oscillator. However, there is a subset of coherent states which may be placed in one-to-one correspondence with the classical harmonic-oscillator states. The classical-made-quantum oscillator is necessarily an indefinite-metric quantum theory for the simple physical reason that the classical bound states cannot have quantized energy levels. Indefinite-metric quantum theories normally have severe problems of physical interpretation. The present work raises the possibility of a partial resolution of some of these problems through a reinterpretation of an indefinite-metric quantum system as a system composed of a classical subsystem interacting with an essentially quantum subsystem of positive metric.

The classical simple harmonic oscillator of frequency ω has the Hamiltonian

$$\mathcal{H} = \frac{1}{2m}(p_1{}^2 + m^2\omega^2 x_1{}^2)\,, \tag{10}$$

and the motion is described by

$$x_1 = A\sin(\pi t + \delta)\,, \tag{11}$$

where A and δ are constants. To embed this classical system in a quantum-mechanical system we introduce the variables x_2 and p_2, and, using Eq. (8), obtain the quantum Hamiltonian

$$\hat{H} = \frac{1}{m} p_1 p_2 + m\omega^2 x_1 x_2\,. \tag{12}$$

We eliminate constants by defining (for $i = 1, 2$)

$$x_i = \left(\frac{1}{m\omega}\right)^{1/2} Q_i\,, \tag{13}$$

$$p_i = (m\omega)^{1/2} P_i\,, \tag{14}$$

and

$$\hat{H} = H\omega \tag{15}$$

so that

$$H = P_1 P_2 + Q_1 Q_2\,. \tag{16}$$

The raising and lowering operators are defined by

$$a_j = \frac{1}{\sqrt{2}}(Q_j + iP_j)\,, \tag{17}$$

and

$$a_j^\dagger = \frac{1}{\sqrt{2}}(Q_j - iP_j) \tag{18}$$

for $j = 1, 2$. They have the commutation relations

$$[a_i, a_j] = [a_i^\dagger, a_j^\dagger] = 0\,, \tag{19}$$

$$[a_i, a_j^\dagger] = 1 - \delta_{ij} \tag{20}$$

for $i,j=1,2$. As a result H is seen to have the form

$$H=\tfrac{1}{2}(a_1a_2^\dagger+a_2a_1^\dagger+a_1^\dagger a_2+a_2^\dagger a_1)\,. \tag{21}$$

The number operators are defined by

$$N_1=a_2a_1^\dagger \tag{22}$$

and

$$N_2=a_2^\dagger a_1 \tag{23}$$

and are not Hermitian. However, their sum is Hermitian and we see that

$$H=N_1+N_2\,. \tag{24}$$

The number operators have the following commutation relations with the raising and lowering operators:

$$N_ia_j=a_j(N_i+\delta_{ij}-1) \tag{25}$$

and

$$N_ia_j^\dagger=a_j^\dagger(N_i-\delta_{ij}+1) \tag{26}$$

for $i,j=1,2$.

Up to this point we have maintained a symmetry of the dynamics under the exchange of the subscripts, $1 \leftrightarrow 2$. Now we must break that symmetry by choosing a vacuum state which is an eigenstate of Q_1 and P_1 or alternately a_1 and $a_1^\dagger$. The commutativity of Q_1 and P_1 permit this. The observability of Q_1 and P_1 for all time requires it. So we define

$$a_1^\dagger|0\rangle=a_1|0\rangle=0\,. \tag{27}$$

As a result $a_2|0\rangle\neq 0$ and $a_2^\dagger|0\rangle\neq 0$. The eigenstates of the number operators are

$$|n_+,n_-\rangle=(a_2^\dagger)^{n_+}(a_2)^{n_-}|0,0\rangle \tag{28}$$

and satisfy

$$N_1|n_+,n_-\rangle=-n_-|n_+,n_-\rangle\,, \tag{29}$$

$$N_2|n_+,n_-\rangle=n_+|n_+,n_-\rangle\,, \tag{30}$$

so that

$$H|n_+,n_-\rangle=(n_+-n_-)|n_+,n_-\rangle\,. \tag{31}$$

The lack of a lower bound to the energy spectrum is in a sense a problem but a necessary one in that it leads to the possibility of bound states with a continuous energy spectrum—a requirement of a faithful representation of the classical oscillator states. There is a subset of coherent states which can be put in a one-to-one relation with the set of classical oscillator states. The defining property of that subset is that its elements are eigenstates of the operators a_1 and $a_1^\dagger$. If we expand an element of that subset in terms of the number eigenstates

$$|z\rangle=\sum_{n_+,n_-=0}^{\infty}f(z|n_+,n_-)|n_+,n_-\rangle \tag{32}$$

and use

$$a_1^\dagger|n_+,n_-\rangle=-n_-|n_+,n_--1\rangle\,, \tag{33}$$

$$a_1|n_+,n_-\rangle=n_+|n_+-1,n_-\rangle \tag{34}$$

to evaluate the eigenvalue equations

$$a_1|z\rangle=iz^*|z\rangle\,, \tag{35}$$

$$a_1^\dagger|z\rangle=-iz|z\rangle\,, \tag{36}$$

we find

$$f(z|n_+,n_-)=\frac{C(iz^*)^{n_+}(iz)^{n_-}}{n_+!n_-!}\,, \tag{37}$$

where C is a constant. As a result

$$|z\rangle=C\exp[i(za_2+z^*a_2^\dagger)]|0,0\rangle\,. \tag{38}$$

We shall call the $|z\rangle$ states coherent states because of their close formal resemblance to the coherent states used in the study of the classical limit of harmonic oscillators, and of quantum electrodynamics[11] (which were eigenstates of the lowering operator but not of the raising operator).

Since $[H,a_1]=-a_1$, and $[H,a_1^\dagger]=a_1^\dagger$, it is clear that the (x_1,p_1) phase-space trajectory is sharp on the set of coherent $|z\rangle$ states. The classical trajectory represented by the state $|z\rangle$ is easily seen to be

$$x_1=\left(\frac{2}{m\omega}\right)^{1/2}R\sin(\omega t+\delta) \tag{39}$$

and

$$p_1=(2m\omega)^{1/2}R\cos(\omega t+\delta)\,, \tag{40}$$

where $z=Re^{i\delta}$. The linearity of H in the "momenta", $\Pi=(p_2,-x_2)$, is crucial for the observability of the phase-space trajectory. In fact, the linearity of all generators of canonical transformations in the momenta is necessary if the canonical transformations are not to take states out of the subset of coherent states.

The superselection rule which follows from the unobservability of the momenta, Π, is best approached by a consideration of the momentum-and coordinate-space representations of the coherent states. In the coordinate-space representation we find that Eqs. (35) and (36) give

$$\left[\left(\frac{m\omega}{2}\right)^{1/2}x_1+i\left(\frac{1}{2m\omega}\right)^{1/2}p_1\right]\langle x_1p_1|z\rangle=iz^*\langle x_1p_1|z\rangle \tag{41}$$

and

$$\left[\left(\frac{m\omega}{2}\right)^{1/2}x_1-i\left(\frac{1}{2m\omega}\right)^{1/2}p_1\right]\langle x_1p_1|z\rangle=-iz\langle x_1p_1|z\rangle\,, \tag{42}$$

so that

$$\langle x_1 p_1 | z\rangle = \sqrt{2}\,\delta\left(x_1 - \left(\frac{2}{m\omega}\right)^{1/2} \mathrm{Im} z\right) \times \delta(p_1 - (2m\omega)^{1/2}\mathrm{Re} z). \tag{43}$$

We have normalized $\langle x_1 p_1 | z\rangle$ so that

$$\langle z' | z\rangle = \int_{-\infty}^{\infty} dx_1 dp_1 \langle z' | x_1 p_1\rangle\langle x_1 p_1 | z\rangle = \delta(\mathrm{Re} z - \mathrm{Re} z')\delta(\mathrm{Im} z - \mathrm{Im} z'). \tag{44}$$

In momentum space Eqs. (35) and (36) lead to the differential equations

$$\left[\left(\frac{m\omega}{2}\right)^{1/2} i\frac{d}{dp_2} + \left(\frac{1}{2m\omega}\right)^{1/2}\frac{d}{dx_2}\right]\langle x_2 p_2 | z\rangle = iz^*\langle x_2 p_2 | z\rangle \tag{45}$$

and

$$\left[\left(\frac{m\omega}{2}\right)^{1/2} i\frac{d}{dp_2} - \left(\frac{1}{2m\omega}\right)^{1/2}\frac{d}{dx_2}\right]\langle x_2 p_2 | z\rangle = -iz\langle x_2 p_2 | z\rangle. \tag{46}$$

They are easily integrated to give

$$\langle x_2 p_2 | z\rangle = \frac{1}{\sqrt{2\pi}}\exp\left[-ip_2\left(\frac{2}{m\omega}\right)^{1/2}\mathrm{Im} z + ix_2(2m\omega)^{1/2}\mathrm{Re} z\right] \tag{47}$$

with the normalization condition

$$\langle z' | z\rangle = \int_{-\infty}^{\infty} dx_2 dp_2 \langle z' | x_2 p_2\rangle\langle x_2 p_2 | z\rangle = \delta(\mathrm{Re} z - \mathrm{Re} z')\delta(\mathrm{Im} z - \mathrm{Im} z'). \tag{48}$$

The transformation function between the two representations is

$$\langle x_1 p_1 | x_2 p_2\rangle = \frac{1}{2\pi}\exp(+ip_2 x_1 - ip_1 x_2), \tag{49}$$

so that

$$\langle x_1 p_1 | z\rangle = \int_{-\infty}^{\infty} dx_2 dp_2 \langle x_1 p_1 | x_2 p_2\rangle\langle x_2 p_2 | z\rangle. \tag{50}$$

Each coherent state, $|z\rangle$, is a superselection sector in itself. There is no measurable dynamical variable $F = F(a_1, a_1^\dagger)$ which connects different states:

$$\langle z' | F(a_1, a_1^\dagger) | z\rangle = F(iz^*, -iz)\delta^2(z - z'). \tag{51}$$

This reflects the lack of a superposition principle in classical mechanics.

The operator formalism for coherent states is incomplete in that we have not defined an inner product. To remedy this deficiency we define the vacuum dual to $|0,0\rangle$ to satisfy

$$\langle 0,0 | a_2 = \langle 0,0 | a_2^\dagger = 0 \tag{52}$$

with $\langle 0,0 | 0,0\rangle = 1$. The dual state corresponding to the physical state, z, we define to be

$$\langle z | = \langle 0,0 | \delta(ia_1 + z^*)\delta(ia_1^\dagger - z) \equiv \langle 0,0 | \int_{-\infty}^{\infty}\frac{d\alpha d\beta}{(2\pi)^2}\exp[i\alpha(\mathrm{Im} z - 2^{-1/2}Q_1) + i\beta(\mathrm{Re} z - 2^{-1/2}P_1)] \tag{53}$$

so that Eqs. (48) and (51) follow if we choose $C = 1$.

Sometimes the dynamical state of a classical system is incompletely known and one only has a set of probabilities that the system is at a particular phase-space point at $t = 0$. If we let $P(z)$ be the probability that the system is at a phase-space point corresponding to z (as defined above), then using the properties

$$P(z) \geq 0, \quad \int d^2z\, P(z) = 1 \tag{54}$$

one sees that a density operator

$$\rho\delta^2(0) = \int d^2z\, |z\rangle P(z)\langle z| \tag{55}$$

may be defined which satisfies

$$\mathrm{Tr}\rho = 1 \tag{56}$$

and

$$\langle z' | \rho | z'\rangle \equiv \lim_{z'' \to z'}\langle z'' | \rho | z'\rangle = P(z'). \tag{57}$$

The mean value of an observable $A = A(a_1, a_1^\dagger)$ is given by

$$\langle A\rangle = \mathrm{Tr}\rho A = \int d^2z\, A(iz^*, -iz)P(z), \tag{58}$$

and one can develop a formalism similar to the density-matrix formalism of quantum mechanics.

We now turn to a closer investigation of the relation of the pseudoquantum mechanics discussed above and true quantum-mechanical systems. We shall be particularly interested in the relation of the coherent states described above and the coherent states of a quantum-mechanical harmonic oscillator—to which they bear such a remarkable resemblance. We shall see that the pseudoquantum oscillator system is equivalent to an indefinite-metric quantum system composed of a harmonic oscillator (thus the connection to the coherent-state quantum oscillator formalism) and an "inverted" oscillator to be described below.

Let us define the following rotated raising and lowering operators in terms of the operators defined in Eqs. (17) and (18):

$$b_1 = a_1\cos\theta + a_2\sin\theta, \tag{59}$$

$$b_2 = -a_1\sin\theta + a_2\cos\theta. \tag{60}$$

Their commutation relations are

$$[b_1, b_1^\dagger] = \sin(2\theta)\,, \tag{61}$$

$$[b_2, b_2^\dagger] = -\sin(2\theta)\,, \tag{62}$$

$$[b_2, b_1^\dagger] = [b_1, b_2^\dagger] = \cos(2\theta) \tag{63}$$

with all other commutators equal to zero. The Hamiltonian of Eq. (21) becomes

$$H = \tfrac{1}{2}(\{b_1, b_1^\dagger\} - \{b_2, b_2^\dagger\})\sin(2\theta) + \tfrac{1}{2}(\{a_1, a_2^\dagger\} + \{a_2, a_1^\dagger\})\cos(2\theta)\,, \tag{64}$$

where $\{u, v\} = uv + vu$.

Now θ is an arbitrary angle and it is obvious that choosing $\theta = 0$ gives the commutation relations and Hamiltonian studied above. However, the choice $\theta = \pi/4$ results in a new form for H and the commutation relations, which can be interpreted as a harmonic oscillator (the b_1 and $b_1^\dagger$ sector) and an "inverted" harmonic oscillator (the b_2 and $b_2^\dagger$ sector) where the commutator and b_2 terms in the Hamiltonian have the wrong sign. The commutativity of the oscillator raising and lowering operators with the inverted oscillator raising and lowering operators leads to a simple factorization of the coherent states which lays bare the basic of the close similarity of form for our coherent states and the coherent states of a quantum oscillator[10]:

$$|z\rangle = \frac{1}{\sqrt{2\pi}} \exp\left[\frac{i}{\sqrt{2}}(z b_1 + z^* b_1^\dagger)\right] \times \exp\left[\frac{i}{\sqrt{2}}(z b_2 + z^* b_2^\dagger)\right] |0, 0\rangle\,, \tag{65}$$

while the coherent state of Ref. 11 has the form

$$|\alpha\rangle = \exp(\alpha b^\dagger - \alpha^* b)|0\rangle\,, \tag{66}$$

where α is a complex numer and $[b, b^\dagger] = 1$. It should be remembered that our choice of vacuum state such that $a_1|0,0\rangle = a_1^\dagger|0,0\rangle = 0$ obviates a simple direct relationship.

Since we have uncovered an interesting relation between a classical-made-quantum system and a "quantum" system of indefinite metric the possibility of reinterpreting indefinite-metric quantum systems as systems containing classical subsystems naturally arises.

III. EMBEDDING OF CLASSICAL FIELDS

In this section we shall discuss the embedding of a classical field theory in a quantum field theory. We shall study the embedding in detail for a scalar field and then describe the features of a classical-made-quantum electrodynamics which we shall call pseudoquantum electrodynamics for the sake of brevity.

Consider a classical field, $\phi_1(x)$, with canonically conjugate momentum, $\pi_1(x)$, and Hamiltonian equations of motion

$$\frac{d}{dt}\phi_1(x) = \frac{\delta \hat{H}}{\delta \pi_1(x)}\,, \tag{67}$$

$$\frac{d}{dt}\pi_1(x) = \frac{-\delta H}{\delta \phi_1(x)}\,, \tag{68}$$

where $\hat{H}$ is the Hamiltonian. We wish to define a "quantum" Hamiltonian, H, which allows us to rewrite Eqs. (67) and (68) in commutator form:

$$\frac{d}{dt}\phi_1(x) = i[H, \phi_1(x)]\,, \tag{69}$$

$$\frac{d}{dt}\pi_1(x) = i[H, \pi_1(x)]\,. \tag{70}$$

Equations (69) and (70) are satisfied if

$$H = \int d^3x \left[\frac{\delta H}{\delta \pi_1(x)} \frac{1}{i} \frac{\delta}{\delta \phi_1(x)} - \frac{\delta H}{\delta \phi_1(x)} \frac{1}{i} \frac{\delta}{\delta \pi_1(x)}\right]. \tag{71}$$

We now formally define

$$\phi_2(x) = i \frac{\delta}{\delta \pi_1(x)} \tag{72}$$

and

$$\pi_2(x) = -i \frac{\delta}{\delta \phi_1(x)}\,, \tag{73}$$

so that

$$H = \int d^3x \left[\frac{\delta \hat{H}}{\delta \pi_1(x)} \pi_2(x) + \frac{\delta \hat{H}}{\delta \phi_1(x)} \phi_2(x)\right]. \tag{74}$$

The fields satisfy the equal-time commutation relations

$$[\phi_i(x), \pi_j(y)] = i(1 - \delta_{ij})\delta^3(\vec{x} - \vec{y})\,, \tag{75}$$

$$[\phi_i(x), \phi_j(y)] = 0\,, \tag{76}$$

$$[\pi_i(x), \pi_j(y)] = 0\,, \tag{77}$$

where δ_{ij} is the Kronecker δ.

We note that the linearity of H in ϕ_2 and π_2 is necessary to maintain the classical character of ϕ_1 and π_1. This is best seen by an examination of Eqs. (69) and (70) and the corresponding Hamiltonian equations for ϕ_2 and π_2. (Other generators of canonical transformations are also linear in π_2 and ϕ_2.)

$\phi_2(x)$ and $\pi_2(x)$ will not be observables on the set of physical states, so that $\phi_1(x)$ and $\pi_1(x)$ will both be sharp on the set of physical states and satisfy superselection rules.

If we wish to couple the classical field to a truly quantum system and maintain the classical nature of the field then certain restrictions exist on the form of the total Hamiltonian H_{tot} and on the commutation relations of the various terms occurring in it. First, the coupling must satisfy the requirement that H_{tot} is linear in $\phi_2(x)$ and $\pi_2(x)$. If we denote the quantum fields by ψ and write the general form of the Hamiltonian as

$$H_{tot} = H + H_Q(\psi) + H_{int}\,, \tag{78}$$

where H is given by Eq. (74), $H_Q(\psi)$ depends only on the quantum fields, ψ, and

$$H_{int} = \int d^3x[\tilde{A}(\phi_1,\pi_1,\psi)\phi_2(x) + \tilde{B}(\phi_1,\pi_1,\psi)\pi_2(x) + \tilde{C}(\phi_1,\pi_1,\psi)]\,, \tag{79}$$

then we can rearrange the Hamiltonian so that

$$H_{tot} = \int d^3x[A(\phi_1,\pi_1,\psi)\phi_2(x) + B(\phi_1,\pi_1,\psi)\pi_2(x) + C(\phi_1,\pi_1,\psi)]\,, \tag{80}$$

where

$$A = \frac{\delta\hat{H}}{\delta\phi_1(x)} + \tilde{A}\,, \tag{81}$$

$$B = \frac{\delta\hat{H}}{\delta\pi_1(x)} + \tilde{B}\,, \tag{82}$$

and

$$C = \tilde{C} + \mathcal{H}_Q \tag{83}$$

with $H_Q = \int d^3x\,\mathcal{H}_Q$. An examination of the equations of motion of $\phi_1(x)$, $\pi_1,(x)$, and ψ,

$$\frac{d}{dt}\phi_1 = B(\phi_1,\pi_1,\psi)\,, \tag{84}$$

$$\frac{d}{dt}\pi_1 = A(\phi_1,\pi_1,\psi)\,, \tag{85}$$

$$\frac{d}{dt}\psi = i[H_{tot},\psi]\,, \tag{86}$$

and the second time derivatives of ϕ_1 and π_1, such as

$$\frac{d^2}{dt^2}\phi_1(x) = i[H,B] = \int d^3y\left(-A\frac{\delta B}{\delta\pi_1(y)} + B\frac{\delta B}{\delta\phi_1(y)} + i\phi_2(y)[A,B] + i\pi_2(y)[B(y),B(x)] + i[C,B]\right), \tag{87}$$

leads us to require the equal-time commutation relations

$$[A(x),A(y)] = [A(x),B(y)] = [B(x),B(y)] = 0\,, \tag{88}$$

where $A(x) = A(\phi_1(x),\pi_1(x),\psi(x))$, etc., so that $\phi_1(x)$ and $\pi_1(x)$ are independent of ϕ_2 and π_2 and hence observable for all time. An examination of higher time derivatives of ϕ_1 and π_1 lead to further restrictions on the equal-time commutation relations of A, B, and C. Examples are

$$[A,[C,B]] = 0\,, \tag{89}$$

$$[B,[C,B]] = 0\,, \tag{90}$$

$$[A,[C,[C,[C,B]]]] = 0\,, \tag{91}$$

etc. A sufficient condition for satisfying all relations of this class consists of having equal-time commutation relations with the form

$$[A,C] = F_1(A,B,\phi_1,\pi_1) \tag{92}$$

and

$$[B,C] = F_2(A,B,\phi_1,\pi_1)\,. \tag{93}$$

Finally, we note that another obvious requirement [cf. Eqs. (84) and (85)] for the observability of ϕ_1 and π_1 is that A and B depend only on an (equal-time) commutative subset of the quantum field variables, ψ.

The above restrictions on the equal-time commutation relations have a direct interpretation in terms of Feynman diagrams for quantum corrections to the classical field behavior. For example, consider the interaction of the classical field sector with a scalar quantum field, ψ, expressed in the interaction

$$H_{int} = g\phi_2(x)\psi^2(x). \tag{94}$$

If $H_Q(\psi)$ is the conventional free Klein-Gordon Hamiltonian, then we find that Eq. (92) is not satisfied so that the Green's function for the classical ϕ_1 field receives quantum corrections from vacuum polarization loops of ψ particles and thus loses its classical character.

We now define a Lagrangian appropriate to our pseudoquantum field theory and then verify the reasonableness of our definition, and the pseudoquantization procedure described above, by studying the equivalent path-integral formulation. The Lagrangian corresponding to the pseudoquantum Hamiltonian, H, is

$$L = \int d^3x(\pi_1\mathring{\phi}_2 + \pi_2\mathring{\phi}_1) - H\,, \tag{95}$$

where $L = L(\phi_1,\mathring{\phi}_1,\phi_2,\mathring{\phi}_2)$ and

$$\pi_1 = \frac{\delta L}{\delta\mathring{\phi}_2}\,, \tag{96}$$

$$\pi_2 = \frac{\delta L}{\delta \dot{\phi}_1}. \tag{97}$$

The vacuum-vacuum transition amplitude for the field theory corresponding to the H_{tot} of Eq. (78) will be shown to be

$$W = \int \prod_x d\phi_1(x) d\phi_2(x) d\pi_1(x) d\pi_2(x) d\psi(x) \exp(iS), \tag{98}$$

where $S = \int dt\, L_{tot}$ up to external source terms. We begin by considering the vacuum-vacuum transition amplitude corresponding to H_Q,

$$W_Q = \int \prod_x d\psi(x) \exp(iS_Q), \tag{99}$$

where ϕ_1 has the character of an external source.

We can now introduce the classical behavior of the ϕ_1 field through functional δ functions

$$\int \prod_x d\psi(x) d\phi_1(x) d\pi_1(x) \delta(B(\phi_1, \pi_1, \psi) - \dot{\phi}_1) \times \delta(A(\phi_1, \pi_1, \psi) + \dot{\pi}_1) e^{iS_Q}, \tag{100}$$

which can be put in the form

$$\int \prod_x d\phi_1(x) d\pi_1(x) d\phi_2(x) d\pi_2(x) \times \exp\left\{ i \int d^4x [(\dot{\phi}_1 - B)\pi_2 - (\dot{\pi}_1 + A)\phi_2] + iS_Q \right\}. \tag{101}$$

After performing a partial integration on the $\dot{\pi}_1\phi_2$ term and discarding a surface term we see that the definition of L in Eq. (95) is correct and that the vacuum-vacuum transition amplitude is indeed given by Eq. (98).

The restrictions on the commutation relations of the various terms in the H_{tot} [expressed in Eqs. (88)-(93)] translate into the requirement that the "quantum completion"[11] of the ϕ_2 field does not take place, i.e., that all N-point functions of the ϕ_2 field are zero:

$$\frac{\delta^n W}{\delta J_2(x_1) \delta J_2(x_2) \cdots \delta J_2(x_n)} = 0, \tag{102}$$

where J_2 is an external source coupled to ϕ_2.

We now discuss the embedding of a free classical Klein-Gordon field in a quantum field theory. The Lagrangian density is

$$\mathcal{L} = \frac{\partial \phi_1}{\partial x^\mu} \frac{\partial \phi_2}{\partial x_\mu} - m^2 \phi_1 \phi_2, \tag{103}$$

from which one obtains the Euler-Lagrange equations (for $i = 1, 2$)

$$(\Box + m^2)\phi_i(x) = 0. \tag{104}$$

The canonical momenta are (note that π_2 is conjugate to ϕ_1, etc.)

$$\Pi_i = \dot{\phi}_i \tag{105}$$

for $i = 1, 2$ with the equal-time commutation relations given by Eqs. (75)-(77). We expand the fields in Fourier integrals:

$$\phi_1(\vec{x}, t) = \int d^3k [a_1(k) f_k(x) + a_1^\dagger f_k^*(x)] \tag{106}$$

and

$$\phi_2(\vec{x}, t) = \int d^3k [a_2(k) f_k(x) + a_2^\dagger(k) f_k^*(x)], \tag{107}$$

where

$$f_k(x) = (2\pi)^{-3/2} (2\omega_k)^{-1/2} e^{-ik\cdot x} \tag{108}$$

with $\omega_k = (\vec{k}^2 + m^2)^{1/2}$. The Fourier component operators satisfy the commutation relations

$$[a_i(k), a_j^\dagger(k')] = (1 - \delta_{ij}) \delta^3(\vec{k} - \vec{k}') \tag{109}$$

and

$$[a_i(k), a_j(k')] = [a_i^\dagger(k), a_j^\dagger(k')] = 0 \tag{110}$$

for $i, j = 1, 2$.

In terms of the Fourier coefficients

$$H = \int d^3x (\dot{\phi}_1 \dot{\phi}_2 + \vec{\nabla}\phi_1 \cdot \vec{\nabla}\phi_2 + m^2 \phi_1 \phi_2) \tag{111}$$

becomes

$$H = \int d^3k\, \omega_k [\{a_1(k), a_2^\dagger(k)\} + \{a_2(k), a_1^\dagger(k)\}]. \tag{112}$$

The analogy between the mode amplitudes of the fields and the raising and lowering operators of the simple harmonic oscillator has been previously remarked. We can therefore use the considerations of Sec. II to establish the spectrum of physical states. The defining properties of a physical state are that $\phi_1(x)$ and $\pi_1(x)$ are sharp on it for all time:

$$\phi_1(x) |\Phi, \Pi\rangle = \Phi(x) |\Phi, \Pi\rangle \tag{113}$$

and

$$\pi_1(x) |\Phi, \Pi\rangle = \Pi(x) |\Phi, \Pi\rangle, \tag{114}$$

where $\Phi(x)$ and $\Pi(x)$ are c-number functions of x:

$$\Phi(x) = \int d^3k [\alpha(k) f_k(x) + \alpha^*(k) f_k^*(x)] \tag{115}$$

and

$$\Pi(x) = -i \int d^3k\, \omega_k [\alpha(k) f_k(x) - \alpha^*(k) f_k^*(x)] \tag{116}$$

with $\alpha(k)$ a c-number function of k.

As a result we are led to define a set of physical states, $|\alpha\rangle$, which are in one-to-one correspondence with the classical solutions of the Klein-Gordon equation and satisfy

$$a_1(k)|\alpha\rangle = \alpha(k)|\alpha\rangle, \quad (117)$$

$$a_1^\dagger(k)|\alpha\rangle = \alpha^*(k)|\alpha\rangle. \quad (118)$$

In analogy with the states of the simple harmonic oscillator (Sec. II) we further define

$$|\alpha\rangle = C\exp\left\{\int d^3k'[\alpha(k')a_2^\dagger(k') - \alpha^*(k')a_2(k')]\right\}|0\rangle, \quad (119)$$

where the vacuum state, $|0\rangle$, satisfies

$$a_1(k)|0\rangle = a_1^\dagger(k)|0\rangle = 0. \quad (120)$$

The physical states, $|\alpha\rangle$, lie in a space which is the infinite tensor product of single-mode spaces. While ϕ_1 and π_1 are sharp for all time on the subset of physical states, we see that ϕ_2 and π_2 are not and, in fact, when applied to a physical state map it into an unphysical state. The superselection rules are embodied in

$$\langle\alpha'|\mathcal{O}|\alpha\rangle = \mathcal{O}_\alpha\delta^2(\alpha-\alpha'), \quad (121)$$

where $\mathcal{O}$ is the operator corresponding to any observable, $\mathcal{O}_\alpha$ is its eigenvalue for the state $|\alpha\rangle$, and $\delta^2(\alpha-\alpha')$ is a functional δ function in the real and imaginary parts of $\alpha-\alpha'$. The functional δ functions have their origin in the definition of the dual set of physical states. We define the dual vacuum state $\langle 0|$ by

$$\langle 0|a_2(k) = 0 \quad (122a)$$

and

$$\langle 0|a_2^\dagger(k) = 0 \quad (122b)$$

for all k with $\langle 0|0\rangle = 1$. The dual state corresponding to $\alpha(k)$ we define by

$$\langle\alpha| = \langle 0|\prod_k \delta(\alpha(k)-a_1(k))\delta(\alpha^*(k)-a_1^\dagger(k)) \equiv \langle 0|\delta(\alpha-a_1)\delta(\alpha^*-a_1^\dagger), \quad (123)$$

so that

$$\langle\alpha'|\alpha\rangle = \delta^2(\alpha'-\alpha) \quad (124)$$

if $C=1$.

We have now established a procedure for embedding a classical field in a quantum field theory. Given a Lagrangian, L, for a classical field theory describing a field $\phi_1(x)$, the Lagrangian density for the pseudoquantum field theory, $\mathcal{L}_{PQ}$ is

$$\mathcal{L}_{PQ}(\phi_1,\dot\phi_1,\phi_2,\dot\phi_2) = \frac{\delta L}{\delta\phi_1(x)}\phi_2(x) + \frac{\delta L}{\delta\dot\phi_1(x)}\pi_2(x) \quad (125)$$

up to a divergence with

$$\pi_2(x) = \frac{\delta}{\delta\dot\phi_1(x)}\int d^3x\,\mathcal{L}_{PQ}. \quad (126)$$

In the case of a classical electromagnetic field interacting with a quantum electron field, one pseudoquantum model, which describes some electromagnetic processes, has the Lagrangian

$$\mathcal{L} = -\tfrac{1}{2}F^1_{\mu\nu}F^2_{\mu\nu} + \bar\psi(i\nabla\!\!\!/ - eA\!\!\!/_1 - m_0)\psi, \quad (127)$$

where $A^1_\mu(x)$ is the classical electromagnetic field, ψ is the electron field, $A^2_\mu(x)$ is the unobservable auxiliary field, and $F^i_{\mu\nu} = \partial_\nu A^i_\mu - \partial_\nu A^i_\mu$ for $i=1,2$. Although our interpretation of the free electromagnetic part of the Lagrangian, $-\frac{1}{2}F^1_{\mu\nu}F^2_{\mu\nu}$, is new, the actual form of this term appeared some time ago in a generalization of electrodynamics by Mie,[12] and was recently used in an Abelian prototype model for quark confinement.[8] The equations of motion are

$$\partial^\mu F^1_{\mu\nu} = 0, \quad (128)$$

$$\partial^\mu F^2_{\mu\nu} + eJ_\nu = 0, \quad (129)$$

and

$$(i\nabla\!\!\!/ - eA\!\!\!/^1 - m)\psi = 0. \quad (130)$$

The canonical momentum which is conjugate to A^1_μ is

$$\Pi^2_\mu = F^2_{0\mu} \quad (131)$$

and that conjugate to A^2_μ is

$$\Pi^1_\mu = F^1_{0\mu}. \quad (132)$$

We take A^1_μ and Π^1_μ to be classical fields which are observable for all time. A^2_μ and Π^2_μ are not observable. Note that $\mathcal{L}$ is invariant under the independent gauge transformations

$$A^1_\mu \to A^1_\mu + \partial_\mu\Lambda^1(x) \quad (133)$$

and

$$A^2_\mu \to A^2_\mu + \partial_\mu\Lambda^2(x). \quad (134)$$

Since $\Pi^1_0 = \Pi^2_0 = 0$, it is apparent that A^1_0 and A^2_0 are c numbers. If we chose the Coulomb gauge for A^1_μ,

$$\vec\nabla\cdot\vec A^1 = 0, \quad (135)$$

and for A^2_μ,

$$\vec\nabla\cdot\vec A^2 = 0, \quad (136)$$

then we can establish the equal-time commutation relations

$$[\Pi_i^a(\vec{x},t), A_j^b(\vec{y},t)] = i(1-\delta_{ab}) \times \int \frac{d^3k}{(2\pi)^3} e^{i\vec{k}\cdot(\vec{x}-\vec{y})}\left(\delta_{ij} - \frac{k_i k_j}{|\vec{k}|^2}\right) = i(1-\delta_{ab})\delta_{ij}^{tr}(\vec{x}-\vec{y}) \qquad (137)$$

for $a,b=1,2$ and $i,j=1,2,3$.

This pseudoquantum field theory describes the dynamics of quantum electron fields interacting with a free, classical electromagnetic field. A typical perturbation theory matrix element would have the form

$$\langle \mathfrak{A}', 0 | T(\bar{\psi}(x) J^{\mu_1}(x_1) A^1_{\mu_1}(x_1) J^{\mu_2}(x_2) A^1_{\mu_2}(x_2) \cdots J^{\mu_n}(x_n) A^1_{\mu_n}(x_n)\psi(y)) | \mathfrak{A}, 0\rangle, \qquad (138)$$

where $|\mathfrak{A}, 0\rangle$ is the tensor product of an electron vacuum state and an electromagnetic state corresponding to the classical field $\mathfrak{A}_\mu(z)$. Because $A^1_\mu(x)$ is sharp on this state, the matrix element becomes

$$\langle 0 | T(\bar{\psi}(x) J^{\mu_1}(x_1) \cdots J^{\mu_n}(x_n)\psi(y)) | 0\rangle \mathfrak{A}_{\mu_1}(x_1)\mathfrak{A}_{\mu_2}(x_2)\cdots \mathfrak{A}_{\mu_n}(x_n) \qquad (139)$$

modulo a functional δ function in $\mathfrak{A}' - \mathfrak{A}$. Thus this model is equivalent to a quantized electron field interacting with an external electromagnetic field.

Another possibility for a model electrodynamics is realized by letting the interaction term in Eq. (127) above be replaced with

$$L_{int} = -e\bar{\psi}A\!\!\!/_2\psi. \qquad (140)$$

Because the equivalent of the equal-time commutation relation, Eq. (92), is not true in this model, the A^1_μ field loses its purely classical character due to quantum corrections. However, this model may be of value for the study of the modification of the A^1_μ field resulting from the emission of many soft photons by a current.

Since vacuum polarization effects modify the electromagnetic field in this case we define in-field eigenstates (in the transverse gauge) by

$$\vec{A}^1_{in}|\mathfrak{A}\rangle_{in} = \vec{\mathfrak{A}}_{in}|\mathfrak{A}\rangle_{in}, \qquad (141)$$

where

$$|\mathfrak{A}\rangle_{in} = \exp\left[\int d^3k \sum_{\lambda=1}^{2}(\alpha(k,\lambda)a_2^\dagger(k,\lambda) - \alpha^*(k,\lambda)a_2(k,\lambda))\right]|0\rangle \qquad (142)$$

and

$$\vec{\mathfrak{A}}_{in} = \int d^3k \sum_{\lambda=1}^{2}\vec{\epsilon}(k,\lambda)[\alpha(k,\lambda)f_k(x) + \alpha^*(k,\lambda)f_k^*(x)] \qquad (143)$$

with

$$\vec{A}^i_{in} = \int d^3k \sum_{\lambda=1}^{2}\vec{\epsilon}(k,\lambda)[a_i(k,\lambda)f_k(x) + a_i^\dagger(k,\lambda)f_k^*(x)] \qquad (144)$$

for $i=1,2$. The vacuum state is defined by

$$a_1(k,\lambda)|0\rangle = a_1^\dagger(k,\lambda)|0\rangle = 0$$

for all k,λ. The interacting field, $\vec{A}^1$, is apparently not sharp on $|\mathfrak{A}\rangle_{in}$ but is sharp on

$$|\mathfrak{A}\rangle = U^{-1}(t,-\infty)|\mathfrak{A}\rangle_{in}, \qquad (145)$$

where

$$U(t,-\infty) = T\left(\exp\left[-i\int_{-\infty}^{t} d^4x\, H_{int}(A^2_{in},\psi_{in})\right]\right) \qquad (146)$$

because

$$\vec{A}^1(\vec{x},t) = U^{-1}(t,-\infty)\vec{A}^1_{in}(\vec{x},t)U(t,-\infty). \qquad (147)$$

With these preliminaries completed, the study of physical processes within the framework of these models is now possible, although we shall not pursue it in this report.

Before turning to a discussion of non-Abelian gauge field theories, it is worth noting that the choice of vacuum state we have made necessitates a redefinition of normal-ordering. By normal-ordering a Lagrangian term we shall mean that the observable fields (to which we have consistently appended the superscript or subscript one) are to be placed to the right, and unobservable fields, labeled by two, are to be placed to the left. Thus Wick's theorem (with our definition of normal-ordering) becomes in the case of two fields

$$\begin{aligned} T(\phi_{1\,in}(x_1)\phi_{2\,in}(x_2)) &= :\phi_{1\,in}(x_1)\phi_{2\,in}(x_2): + \langle 0|T(\phi_{1\,in}(x_1)\phi_{2\,in}(x_2))|0\rangle \\ &= \phi_{2\,in}(x_2)\phi_{1\,in}(x_1) + \theta(x_{10}-x_{20})[\phi_{1\,in}(x_1),\phi_{2\,in}(x_2)]. \end{aligned} \qquad (148)$$

Note that the Green's function

$$G(x_1,x_2) = \langle 0|T(\phi_{1\,in}(x_1)\phi_{2\,in}(x_2))|0\rangle \qquad (149)$$

is necessarily retarded. From this we can conclude that the models of electrodynamics, which we have considered, naturally embody the observed retarded nature of classical electrodynamics. Another way of stating this result is: If classical electrodynamics is to have a pseudoquantum formulation, its Green's functions are necessarily retarded. The origin of the asymmetry is the definition of the vacuums (which is equivalent to a specification of boundary conditions). Just as in classical electrodynamics retarded propagation is implemented by a choice of boundary conditions which do not require a commitment to any specific cosmological model.

Finally we would like to note that the Lagrangian obtained from adding $L_{\rm int}$ of Eq. (140) to the Lagrangian of Eq. (127) is equivalent to the usual Lagrangian of electrodynamics plus a term describing a massless Abelian gauge field with the wrong sign. (This is seen by defining new fields equal to the sum and difference of A_μ^1 and A_μ^2.) This field theory may be quantized following the procedure we have outlined. A_μ^1 loses its classical character due to quantum corrections.

IV. NON-ABELIAN GAUGE THEORIES

In this section we shall describe the procedure for embedding a classical non-Abelian Yang-Mills field in a quantum field theory. Then we will discuss a vierbein formulation of quantum gravity which could have been interpreted as a pseudoquantum field theory for a classical metric field if it were not for one term in the Lagrangian which makes it a truly quantum field theory. Nevertheless we suggest a new canonical quantization procedure based on our pseudoquantum approach.

Consider a classical Yang-Mills field, $A_\mu^1 = \underline{A}_\mu^1 \cdot \underline{T}$, where the jth component of $\underline{T}$ is a matrix representing a generator of a non-Abelian group G in the defining representation with commutation relations

$$[T_j, T_k] = it_{jkl}T_l. \tag{150}$$

We can define a pseudoquantum field theory, wherein the classical character of A_μ^1 is maintained, which has the Lagrangian density

$$\begin{aligned}\mathcal{L} = {} & \tfrac{1}{2}\underline{F}_{\mu\nu}^1 \cdot \underline{F}^{2\mu\nu} - \tfrac{1}{2}\underline{F}^{2\mu\nu}\cdot(\partial_\mu \underline{A}_\nu^1 - \partial_\nu \underline{A}_\mu^1 + g\underline{A}_\mu^1 \times \underline{A}_\nu^1) \\ & -\tfrac{1}{2}\underline{F}^{1\mu\nu}\cdot(\partial_\mu \underline{A}_\nu^2 - \partial_\nu \underline{A}_\mu^2 + g\underline{A}_\mu^1\times \underline{A}_\nu^2 - g\underline{A}_\nu^1 \times \underline{A}_\mu^2) \\ & + \bar{\psi}(i\not\nabla + g\not A^1 - m)\psi,\end{aligned} \tag{151}$$

where ψ is a fermion field. The theory is invariant under the local gauge transformation, $S \in G$,

$$\psi' = S^{-1}\psi, \tag{152}$$

$$A_\mu^{1'} = S^{-1}A_\mu^1 S + \frac{i}{g}S^{-1}\partial_\mu S\,, \tag{153}$$

$$F_{\mu\nu}^{1'} = S^{-1}F_{\mu\nu}^1 S, \tag{154}$$

$$A_\mu^{2'} = S^{-1}A_\mu^2 S, \tag{155}$$

$$F_{\mu\nu}^{2'} = S^{-1}F_{\mu\nu}^2 S. \tag{156}$$

Except for one important term this Lagrangian with its attendant gauge invariance properties has been suggested as a possible model for the quark-confining strong interaction.[8] Since the omitted term has a masslike character $\Lambda^2 \underline{A}_\mu^2 \cdot \underline{A}^{2\mu}$, where Λ has the dimensions of a mass, it is clear that the strong-interaction model's ultraviolet behavior approaches that of the present pseudoquantum theory if the same quantization procedure is followed in both cases. We shall discuss this question further in the next section and show that the *ad hoc* procedure followed in Ref. 8 leads to the same result as the quantization procedure developed in this report.

The Euler-Lagrange equations of motion which are obtained from $\mathcal{L}$ in the canonical manner are

$$\underline{F}_{\mu\nu}^1 = \partial_\mu \underline{A}_\nu^1 - \partial_\nu \underline{A}_\mu^1 + g\,\underline{A}_\mu^1 \times \underline{A}_\nu^1, \tag{157}$$

$$\underline{F}_{\mu\nu}^2 = \partial_\mu \underline{A}_\nu^2 - \partial_\nu \underline{A}_\mu^2 + g\underline{A}_\mu^1 \times \underline{A}_\nu^2 - g\underline{A}_\nu^1 \times \underline{A}_\mu^2, \tag{158}$$

$$(\partial_\mu + g\,\underline{A}_\mu^1\times)\underline{F}^{1\mu\nu} = 0, \tag{159}$$

$$(\partial_\mu + g\underline{A}_\mu^1\times)\underline{F}^{2\mu\nu} + g\underline{A}_\mu^2 \times \underline{F}^{1\mu\nu} + g\underline{J}^\nu = 0, \tag{160}$$

$$(i\not\nabla + g\not A^1 - m)\psi = 0, \tag{161}$$

with the conservation law

$$(\partial_\nu + g\underline{A}_\nu^1\times)\underline{J}^\nu = 0. \tag{162}$$

The canonical momentum which is conjugate to $\underline{A}_j^1$ is

$$\underline{\Pi}_j^2 = \underline{F}_{0j}^2 \tag{163}$$

and the canonical momentum conjugate to $\underline{A}_j^2$ is

$$\underline{\Pi}_j^1 = \underline{F}_{0j}^1 \tag{164}$$

for $j = 1, 2, 3$. The canonical momentum corresponding to the fields $\underline{A}_0^i$ is zero for $i = 1, 2$. The existence of equations of constraint among the Euler-Lagrange equations implies that not all field components are independent, so that we must isolate the independent components prior to defining the canonical equal-time commutation relations.

Following Ref. 8 we choose to work in the Coulomb gauge, $\nabla_i A_i^1 = 0$, and define the field variables

$$\underline{A}_i^2 = \underline{A}_i^{2T} + \underline{A}_i^{2L}, \tag{165}$$

$$\underline{\Pi}_i^a = \underline{\Pi}_i^{aT} + \underline{\Pi}_i^{aL}, \tag{166}$$

where

$$\nabla_i \cdot \underline{A}_i^{2T} = \nabla_i \cdot \underline{\Pi}_i^{aT} = 0 \tag{167}$$

and $a = 1, 2$. Then the nonzero equal-time commutation relations are

$$[\Pi_{ip}^{aT}(x), A_{jq}^{bT}(y)] = i\delta_{pq}(1-\delta_{ab})\delta_{ij}^{tr}(\vec{x}-\vec{y}), \tag{168}$$

where p and q are internal-symmetry indices, $a, b = 1, 2$, and $i, j = 1, 2, 3$.

While the classical character of $\underline{A}^1_\mu$ can be maintained with our choice of $\mathcal{L}$, this theory has features due to its non-Abelian nature which make it less trivial and therefore more interesting than the corresponding Abelian theory discussed in the last section. If we follow a procedure similar to that in the Abelian case [Eq. (127)] and introduce a set of states appropriate to the quadratic part of the Lagrangian, then the cubic and quartic Yang-Mills terms in the interaction part of the Lagrangian will act to transform $\underline{A}^1_{\text{in}\,\mu}$ eigenstates into eigenstates of the interacting field $\underline{A}^1_\mu$. This is, of course, necessary for the classical Yang-Mills equations of motion to be satisfied. Our formalism, thus, offers a perturbative method for calculating solutions of the classical Yang-Mills equations. In addition, it gives an interesting interpretation to the short-distance behavior of the quark-confining field theory of Ref. 8. At short distances the gluon field $\underline{A}^1_\mu$ effectively decouples from the quark sector and becomes, in effect, a free field. This type of short-distance behavior is certainly not at odds with the seemingly simple behavior observed in hadron processes at high energy. Therefore, it is possible that pseudoquantum field theory may be relevant to the short-distance behavior of hadron interaction. Certainly, it is interesting that elementary fermions fall into two similar groups: those which appear to be individually observable (leptons) and those which are not individually observable (quarks).

We now turn to a consideration of a vierbein model of gravity which has certain close similarities to the pseudoquantum field theories we have been studying. In Weyl's formulation[13] of the Einstein-Cartan theory of gravity a vierbein field, $l^{\mu a}(x)$, is introduced which is the "square root" of the metric tensor

$$g^{\mu\nu} = \eta_{ab} l^{\mu a} l^{\nu b}, \tag{169}$$

where η_{ab} is the constant metric tensor of special relativity, where Roman indices transform as vectors under the SL(2, C) group of local Lorentz transformations, and where Greek indices transform as vectors under general coordinate transformations. It is useful to introduce the constant Dirac matrices, γ_a and $4S_{ab} = i[\gamma_a, \gamma_b]$. Under an SL(2, C) transformation,

$$S = \exp[iC^{ab}(x)S_{ab}], \tag{170}$$

a spinor, $\psi(x)$, becomes

$$\psi' = S\psi. \tag{171}$$

The local nature of the transformation requires the introduction of a gauge field

$$B^{ab}_\mu = -B^{ba}_\mu \tag{172}$$

which transforms inhomogeneously,

$$B_\mu \to SB_\mu S^{-1} - \frac{i}{g} S\partial_\mu S^{-1}, \tag{173}$$

so that a Lorentz transformation gauge-covariant derivative can be defined

$$\nabla_\mu \psi = (\partial_\mu + igB_\mu)\psi, \tag{174}$$

where $B_\mu = B^{ab}_\mu S_{ab}$ and $g = 12\pi G$ where G is Newton's constant. Under a gauge transformation we have

$$l^\mu = l^{\mu a}\gamma_a \to Sl^\mu S^{-1}, \tag{175}$$

so that the gauge-covariant derivative of l^μ is defined to be

$$\nabla_\nu l^\mu = (\partial_\nu + igB_\nu \times) l^\mu, \tag{176}$$

where $B_\nu \times l^\mu = [B_\nu, l^\mu]$. The commutator

$$igB_{\mu\nu} = [\partial_\mu + igB_\mu, \partial_\nu + igB_\nu] \tag{177}$$

transforms homogeneously under a gauge transformation

$$B_{\mu\nu} \to SB_{\mu\nu}S^{-1}, \tag{178}$$

and as a second-rank tensor under general coordinate transformations. With these field quantities we are able to construct a Lagrangian $\mathcal{L}_{\text{Weyl}}$ which reduces to the Einstein Lagrangian for gravity when no matter is present,[13]

$$\mathcal{L} = \mathcal{L}_{\text{Weyl}} + \mathcal{L}_{\text{matter}}, \tag{179}$$

where

$$\mathcal{L}_{\text{Weyl}} = \frac{i}{8l} \operatorname{Tr} l^\mu l^\nu B_{\mu\nu} \tag{180}$$

and where, for example, we might let

$$l\,\mathcal{L}_{\text{matter}} = \bar{\psi}(il^\mu \nabla_\mu + m)\psi \tag{181}$$

with $l = \det(l^{\mu a})$.

We observe that the terms containing derivatives in $\mathcal{L}_{\text{Weyl}}$ are linear in the field B_μ—a suggestive feature in view of our previous discussion. However, the quadratic term in B_μ eliminates the possibility of regarding $\mathcal{L}_{\text{Weyl}}$ as a pseudoquantum field theory for a classical field $l^{\mu a}$. But, regardless of this consideration, the fact that $l^{\mu a}$ is necessarily classical in part leads us to consider quantizing vierbein gravity in a manner which is based on the pseudoquantization procedure described above. Remembering that a successful perturbation theory requires the perturbation to be around known solutions we introduce a quadratic Lagrangian term via

$$\mathcal{L} = \mathcal{L}_0 + (\mathcal{L} - \mathcal{L}_0) = \mathcal{L}_0 + \mathcal{L}_{\text{int}}, \tag{182}$$

where

$$\mathcal{L}_0 = -\tfrac{1}{4} i \operatorname{Tr}(B'_{\mu a} l^\mu \gamma^a + ig[B_a, B_b]\gamma^a\gamma^b) \tag{183}$$

and

$$B'_{\mu a}=\partial_\mu B_a-\partial_a B_\mu\,. \tag{184}$$

Our plan is to follow the pseudoquantization procedure for the "free" part of the Lagrangian $\mathcal{L}_0$. Therefore we will (i) choose a particular coordinate system (harmonic coordinates) and a particular gauge, the "Lorentz" gauge, $\partial^\mu B_\mu=0$, (ii) establish equal-time commutation relations, (iii) define a set of eigenstates of $l^{\mu a}$, and (iv) proceed to calculate quantum corrections in perturbation theory.

The equations of motion for the "free" Lagrangian $\mathcal{L}_0$ are

$$\partial_\mu B_b^{ab}-\partial_b B_\mu^{ab}=0 \tag{185}$$

and

$$\partial_\mu(l^{\mu a}\eta^{\nu b}-l^{\nu a}\eta^{\mu b})+2g\,(\eta^{\nu a}B_c^{cb}-\eta^{\nu b}B_c^{ca}-\eta^{ac}B_c^{\nu b}+\eta^{bc}B_c^{\nu a})=0\,. \tag{186}$$

We work in the gravitational equivalent of the Lorentz gauge of electrodynamics,

$$\partial^\mu B_\mu^{ab}=0\,, \tag{187}$$

and choose harmonic coordinates

$$\partial_\mu l^{\mu a}=\tfrac{1}{2}\,\partial^a\eta_{\sigma\tau}l^{\sigma\tau}\,. \tag{188}$$

The Green's function associated with Eq. (185) is

$$G_{\alpha ef,\rho\sigma}(x,y)=-\tfrac{1}{2}\int\frac{d^4k}{k^2}\,e^{-ik\cdot(x-y)}g_{\alpha ef,\rho\sigma}(k)\,, \tag{189}$$

where

$$g_{\alpha ef,\rho\sigma}(k)=k_e\left(\eta_{\alpha\rho}\eta_{f\sigma}+\eta_{\alpha\sigma}\eta_{f\rho}-\eta_{\alpha f}\eta_{\rho\sigma}-\frac{k_\alpha k_\rho\eta_{f\sigma}+k_\alpha k_\sigma\eta_{f\rho}}{k^2}\right)-k_f\left(\eta_{\alpha\rho}\eta_{e\sigma}+\eta_{\alpha\sigma}\eta_{e\rho}-\eta_{\alpha e}\eta_{\rho\sigma}-\frac{k_\alpha k_\rho\eta_{e\sigma}+k_\alpha k_\sigma\eta_{e\rho}}{k^2}\right). \tag{190}$$

In order to relate the above Green's function to a time-ordered product of the quantum fields it is first necessary to introduce a set of coherent states, $|L\rangle$, which are eigenstates of $l^{\mu a}$:

$$l^{\mu a}(x)\,|L\rangle=L^{\mu a}(x)\,|L\rangle\,, \tag{191}$$

where $L^{\mu a}(x)$ is a c-number function of x. In particular, we define $|\eta\rangle$ to satisfy

$$l^{\mu a}|\eta\rangle=\eta^{\mu a}|\eta\rangle\,, \tag{192}$$

where $\eta^{\mu a}$ is the constant Lorentz metric tensor of special relativity. Given a state $|L\rangle$ we define the field

$$l_L^{\mu a}=l^{\mu a}-L^{\mu a}\,. \tag{193}$$

This field corresponds to the quantum part of $l^{\mu a}$ and when applied to the purely classical state $|L\rangle$ has the eigenvalue zero.

We now make the identification

$$iG_{\alpha ef,\rho\sigma}(x,y)=\langle L\,|\,T(B_{\alpha ef}(x),l_{L\rho\sigma}(y))\,|\,L\rangle\,. \tag{194}$$

If we desire to calculate quantum corrections to $l_{\rho\sigma}=\eta_{\rho\sigma}$ we choose $|L\rangle=|\eta\rangle$. (It should be noted that $G_{\alpha ef,\rho\sigma}$ is independent of the choice of $|L\rangle$ as we have defined it.) Because $l_{L\rho\sigma}(y)$ is sharp on $|L\rangle$ we find that the right side of Eq. (194) becomes

$$iG_{\alpha ef,\rho\sigma}(x,y)=\theta(y_0-x_0)[l_{\rho\sigma}(y),B_{\alpha ef}(x)] \tag{195}$$

up to a functional δ function. From the form of $\mathcal{L}_0$ we see that the commutator is not zero. It is fully determined by an equal-time commutation relation of $l_{\rho\sigma}$ and $B_{\alpha ef}$ (which by the way is the only nonzero equal-time commutator if the canonical procedure is followed), the equations of motion, and the requirement that it be zero at spacelike distances. The "retarded" form of $G_{\alpha ef,\rho\sigma}$ fixes the integration contour around poles in Eq. (192). The other nonzero Green's function in the free Lagrangian model specified by $\mathcal{L}_0$ is

$$iH^{\mu\nu,\rho\sigma}(x,y)=\langle L\,|\,T(l_L^{\mu\nu}(x),l_L^{\rho\sigma}(y))\,|\,L\rangle\,. \tag{196}$$

It is nonzero owing to the presence of the $[B_\mu,B_\nu]$ term in $\mathcal{L}_0$. We shall show in the next section that it is a principal-value propagator rather than a Feynman propagator. In coordinate space this results in $H^{\mu\nu,\rho\sigma}$ being the sum of the advanced and retarded propagators. As a result our model is equivalent to an action-at-a-distance theory in some sectors.

The classical part of $l_{\mu a}$ is the solution of the classical linearized field equations with appropriate matter sources. The linearized field equations are derived from a Lagrangian consisting of $\mathcal{L}_0$ plus matter terms. (Note that the form of $\mathcal{L}_0$ is obtained by substituting $l_{\mu a}=\eta_{\mu a}+h_{\mu a}$ in $\mathcal{L}_{\rm Weyl}$, expanding, and keeping quadratic terms.) Thus the class of possible background metrics is restricted.

A simplification occurs in perturbation theory when the classical part of $l_{\mu\alpha}$ is $\eta_{\mu\alpha}$. In this case $(\mathcal{L}_{\rm Weyl}-\mathcal{L}_0)|\eta\rangle=0$ when $\mathcal{L}_0$ and $\mathcal{L}_{\rm Weyl}$ are expressed in terms of asymptotic fields.

V. PRINCIPAL-VALUE PROPAGATORS AND ACTION AT A DISTANCE

In this section we shall show that certain propagators, in field theories where the pseudoquantization procedure has been followed, are principal-value propagators (i.e., the sum of the advanced and retarded Green's functions in coordinate space) rather than Feynman propagators. We also describe a quantum field theory for action-at-a-distance electrodynamics which completes the program initiated by Schwarzschild, Tetrode, and Fokker.[14]

To illustrate the origin of the principal-value propagator we return to the scalar field model of Eq. (103) which described a classical field, $\phi_1(x)$. We introduce an interaction term

$$L_{\text{int}} = -\int d^3z\,\tfrac{1}{2}\lambda^2\,[\phi_2(z)]^2 \tag{197}$$

(where λ is a constant), which destroys the purely classical nature of ϕ_1. Suppose we consider the Green's function

$$i\bar{G}(x,y) = \langle 0|\,T(\phi_1(x)\phi_1(y))\,|0\rangle\,, \tag{198}$$

which would be zero if L_{int} were not present. In terms of in-fields we have

$$i\bar{G}(x,y) = \left\langle 0\left|T\left(\phi_{1\text{in}}(x)\phi_{1\text{in}}(y)\exp\left(i\int dt\,L_{\text{int}}\right)\right)\right|0\right\rangle, \tag{199}$$

where the vacuum states, $|0\rangle$ and $\langle 0|$, are defined as in Eqs. (120) and (122). From the definition of the vacuum we find (dropping "in" labels)

$$i\bar{G}(x,y) = \frac{-i\lambda^2}{2}\int d^4z\langle 0|\,T(\phi_1(x)\phi_1(y)\phi_2{}^2(z))\,|0\rangle\,, \tag{200}$$

which becomes

$$i\bar{G}(x,y) = \frac{-i\lambda^2}{2}\,\epsilon(x_0-y_0)\,\frac{\partial}{\partial m^2}\,\Delta(x-y) \tag{201}$$

with

$$\Delta(x-y) = -i\int\frac{d^4k}{(2\pi)^3}\,\delta(k^2-m^2)\epsilon(k_0)e^{-ik\cdot(x-y)}\,. \tag{202}$$

Using

$$\tfrac{1}{2}\epsilon(x_0-y_0)\Delta(x-y) = \int\frac{d^4k}{(2\pi)^4}\,\mathrm{P}\,\frac{1}{k^2-m^2}\times e^{-ik\cdot(x-y)}\,, \tag{203}$$

we see that

$$\bar{G}(x,y) = -\lambda^2\int\frac{d^4k}{(2\pi)^4}\,\mathrm{P}\,\frac{1}{(k^2-m^2)^2}\,e^{-ik\cdot(x-y)}\,, \tag{204}$$

where

$$\mathrm{P}\,\frac{1}{(k^2-m^2)^2} \equiv \frac{1}{2}\left[\frac{1}{(k^2-m^2+i\epsilon)^2}+\frac{1}{(k^2-m^2-i\epsilon)^2}\right]. \tag{205}$$

The form of $\bar{G}$ is consistent with the equations of motion:

$$(\Box+m^2)\phi_1+\lambda^2\phi_2=0\,, \tag{206}$$

$$(\Box+m^2)\phi_2=\delta^4(x-y)\,. \tag{207}$$

The appearance of the principal-value dipole propagator rather than the Feynman dipole propagator in Eq. (204) is useful because it eliminates certain unitarity problems associated with indefinite-metric fields. However, depending on the model under consideration, it could lead to difficulties with causality. To illustrate the manner in which unitarity problems are resolved, consider the interaction of the ϕ_1 dipole field with a scalar quantum field ψ with

$$L'_{\text{int}} = g\phi_1(x)[\psi(x)]^2\,. \tag{208}$$

Suppose we consider the subset of in and out states containing arbitrary numbers of ψ particles but no ϕ_1 or ϕ_2 particles. These states have positive metric. If one could systematically exclude indefinite-metric ϕ_1 and ϕ_2 particles from physical states one would avoid negative probabilities and other problems. But the sum over states in a unitarity sum would normally include states with ϕ_1 particles if the ϕ_1 field had Feynman propagators. In the case of principal-value propagators, no intermediate states with ϕ_1 particles occur, since the pole term is not present. The interaction mediated by the ϕ_1 field is a form of action at a distance and ϕ_1 is properly described by the phrase adjunct field, coined by Feynman and Wheeler.[14] A more detailed discussion of the unitarity question is given in Refs. 7 and 8. In those articles a dipole gluon model for quark confinement was proposed which introduced principal-value propagators in an *ad hoc* manner to resolve unitarity problems. It was pointed out that causality problems did not necessarily exist in those models because the non-Abelian dipole gluons were confined for the same reason as the quarks so that—at the worst—there would be unobservable causality violations at distances of the order of hadron dimensions.

The pseudoquantization procedure may be used to construct a quantum field-theoretic version of action-at-a-distance electrodynamics. Consider the Lagrangian

$$\mathcal{L} = -\tfrac{1}{2}F^{\mu\nu}(\partial_\nu A_\mu-\partial_\mu A_\nu)+\tfrac{1}{4}F^{\mu\nu}F_{\mu\nu} + \bar{\psi}(i\not{\partial}-e\not{A}-m_0)\psi\,. \tag{209}$$

We define the momentum

$$\Pi_\mu = \frac{\delta \mathfrak{L}}{\delta \dot{A}^\mu} = F_{0\mu} . \tag{210}$$

Going to the transverse gauge as in Sec. IV, we define the equal-time commutation relation

$$[\Pi_i(\vec{x}, t), A_j(\vec{y}, t)] = i\delta_{ij}^{tr}(\vec{x} - \vec{y}) . \tag{211}$$

Suppose we neglect interaction terms in $\mathfrak{L}$ for the moment and choose $F_{\mu\nu}$ to be an observable classical field (as it is up to quantum corrections which we neglect) and A_μ to be unobservable (as it is because it is not gauge invariant). Then we follow our pseudoquantization procedure for

$$\mathfrak{L}_0 = -\tfrac{1}{2} F^{\mu\nu}(\partial_\nu A_\mu - \partial_\mu A_\nu) + \tfrac{1}{4} F^{\mu\nu} F_{\mu\nu} . \tag{212}$$

In particular, we define a vacuum such that

$$F_{\mu\nu}|0\rangle = 0, \quad A_\mu|0\rangle \neq 0 , \tag{213}$$

while

$$\langle 0|A_\mu = 0, \quad \langle 0|F_{\mu\nu} \neq 0 . \tag{214}$$

Then

$$iG_{\mu\nu}(x, y) = \langle 0|T(A_\mu(x)A_\nu(y))|0\rangle \tag{215}$$

would be zero were it not for $F_{\mu\nu}F^{\mu\nu}$ in $\mathfrak{L}_0$. In terms of appropriate in-fields it becomes

$$2iG_{\mu\nu}(x, y) = \int d^4z \, (\theta(x_0 - y_0)\theta(y_0 - z_0) + \theta(y_0 - x_0)\theta(x_0 - z_0)) \times [A_{\mu\,\mathrm{in}}(x), F_{\alpha\beta\,\mathrm{in}}(z)][A_{\mu\,\mathrm{in}}(y), F_{\mathrm{in}}^{\alpha\beta}(z)] . \tag{216}$$

Note that we are treating $F_{\mu\nu}F^{\mu\nu}$ in $\mathfrak{L}_0$ as an interaction term. The structure of $G_{\mu\nu}(x, y)$ is the same as that of Eq. (200) so we can conclude that

$$G_{\mu\nu}(x, y) = -g_{\mu\nu} \int \frac{d^4k}{(2\pi)^4} \, \mathrm{P} \frac{1}{k^2} e^{-ik\cdot(x-y)} \tag{217}$$

in the Feynman gauge. Thus the action-at-a-distance interaction follows from the pseudoquantization of electrodynamics. The classical character of $F_{\mu\nu}$ is lost owing to quantum corrections resulting from the presence of $J_\mu A^\mu$ in the Lagrangian.

The example we have just studied has a certain parallel in the vierbein model of gravitation studied in the last section. The forms of the Lagrangian and commutation relations are similar. As a result it is clear that

$$D^{\mu\nu,\lambda\sigma}(x, y) \equiv \left\langle L \left| T\left(l_{L\,\mathrm{in}}^{\mu\nu}(x) l_{L\,\mathrm{in}}^{\lambda\sigma}(y) \int d^4z \, \tilde{\mathfrak{L}}_{\mathrm{int}}(z)\right) \right| L \right\rangle \tag{218}$$

with

$$\tilde{\mathfrak{L}}_{\mathrm{int}} = \tfrac{1}{4} g \, \mathrm{Tr}\, [B_{\mu\,\mathrm{in}}, B_{\nu\,\mathrm{in}}] \gamma^\mu \gamma^\nu \tag{219}$$

is a principal-value propagator. Therefore we have constructed an action-at-a-distance version of quantum gravity. Our motivation was to take account of the classical part of $l^{\mu a}$ in a way which did not divorce it from the quantum part to which it is intimately related.

VI. CONCLUSION

We have seen that an alternative to Fock-space quantization exists for a class of field theories which have Lagrangian gradient terms which are linear in field variables. A method was also proposed for constructing Lagrangians of that type from classical Lagrangians with gradient terms which are quadratic in field variables. To some extent this process has a parallel in the passage from Klein-Gordon field Lagrangians which are quadratic in derivatives to Dirac field Lagrangians which are linear in derivatives.

The quantization procedure we have outlined is canonical so far as the fields are concerned. We do, however, make a choice of vacuum states which differs from the usual choice. As a result we have found free propagators which were either retarded, or half-advanced and half-retarded. The choice of vacuum state does not in itself preclude the appearance of Feynman propagators. If one has a good reason to modify the canonical commutation relations then it is possible to obtain Feynman propagators.[15] The procedure we have outlined has, therefore, a greater generality than the particular class of models studied in the present work. It can enable one to embed a classical field theory in a quantum field theory in such a way as to maintain its classical character. It can also be applied to study classical field theories which obtain quantum corrections. Finally it can be applied in order to obtain a fully second-quantized field theory (cf. Ref. 15).

ACKNOWLEDGMENT

This work was supported in part by the U.S. Energy Research and Development Administration.

*Present address: Physics Department, Williams College, Williamstown, Mass. 01267.

[1]D. R. Yennie, S. C. Frautschi, and H. Suura, Ann. Phys. (N.Y.) 13, 379 (1961).

[2]R. J. Glauber, Phys. Rev. 131, 2766 (1963).

[3]W. A. Bardeen, M. S. Chanowitz, S. D. Drell, M. Weinstein, and T.-M. Yan, Phys. Rev. D 11, 1094 (1975).

[4]J. M. Cornwall and G. Tiktopoulos, Phys. Rev. D 13, 3370 (1976).
[5]S. Blaha, Phys. Lett. 56B, 373 (1975).
[6]E. C. G. Sudarshan, Center for Particle Theory report Univ. of Texas—Austin, 1976 (unpublished).
[7]S. Blaha, Phys. Rev. D 10, 4268 (1974).
[8]S. Blaha, Phys. Rev. D 11, 2921 (1975).
[9]S. Blaha, Lett. Nuovo Cimento 18, 60 (1977).
[10]Cf. Ref. 2; T. W. B. Kibble, J. Math. Phys. 9, 315 (1968); Phys. Rev. 173, 1527 (1968); 174, 1882 (1968); 175, 1624 (1968);
[11]A. Salam, lecture at Center for Theoretical Studies, Miami, Florida, 1973 (unpublished).
[12]G. Mie, Ann. Phys. (Leipzig) 37, 511 (1912); 39, 1 (1912); 40, 1 (1913); H. Weyl, *Space, Time, Matter* (Dover, N.Y. 1952).
[13]H. Weyl, Z. Phys. 56, 330 (1929); T. W. B. Kibble, J. Math. Phys. 2, 212 (1961); J. Schwinger, Phys. Rev. 130, 1253 (1963); C. J. Isham, A. Salam, and J. Strathdee, Lett. Nuovo Cimento 5, 969 (1972); F. W. Hehl, P. von der Heyde, G. D. Kerlick, and J. Nester, Rev. Mod. Phys. 48, 393 (1976); and references therein.
[14]K. Schwarzschild, Göttinger Nachrichten 128, 132 (1903); H. Tetrode, Z. Phys. 10, 317 (1922); A. D. Fokker, *ibid.* 58, 386 (1929); J. Wheeler and R. P. Feynman, Rev. Mod. Phys. 17, 157 (1945); 21, 425 (1949).
[15]S. Blaha (unpublished).

Appendix D. PseudoQuantum Non-Abelian Field Theory Paper

This refereed paper is S. Blaha, Phys. Rev. D**11**, 2921 (1975). Reprinted with the kind permission of Physical Review D.

PHYSICAL REVIEW D VOLUME 11, NUMBER 10 15 MAY 1975

Second-quantized non-Abelian field theory for hadrons with quark confinement and scaling deep-inelastic structure functions*

Stephen Blaha
Laboratory of Nuclear Studies, Cornell University, Ithaca, New York 14853
(Received 30 December 1974)

A four-dimensional second-quantized field theory with quarks bound by "colored" non-Abelian gluons is described which has the following properties: (1) the only physical particles are color singlets composed solely of quarks, (2) the deep-inelastic structure functions have Bjorken scaling, (3) gluon loops and Faddeev-Popov ghost loops are identically zero in any gauge, (4) Regge trajectories are apparently linear on a Chew-Frautschi plot, and (5) constituent motion within hadrons can be nonrelativistic.

I. INTRODUCTION

After a period of some skepticism the possibility that hadronic interactions might be understood within the framework of quantum field theory is again being seriously considered.[1] This is partly the result of the psychological climate created by the apparently successful unification of weak and electromagnetic interactions in a renormalizable field theory and partly the result of a greater appreciation of the variety of phenomena which can occur in field theories.

In this article we shall describe a field-theoretic model of hadron binding which has two major features: (1) Hadrons only occur as quark-antiquark or three-quark bound states, and (2) quarks behave as quasifree particles within hadrons. We assume that the suggestions of an internal symmetry called color[2] are correct and that the strong interaction consists of the exchange of colored Yang-Mills gluons. The nature of the interaction allows only color singlet states to occur in the gauge-invariant physical particle spectrum and consequently the first feature will be realized by choosing the color group to be SU(3). Since the (Schwinger) mechanism which produces this result is an infrared phenomenon, the second feature is not precluded and the model is essentially free in the ultraviolet region of the quark sector.

Our model is a non-Abelian version of a recently investigated Abelian field theory which had quark confinement and scaling electroproduction structure functions.[3] In that theory the free propagator of the massless gluon field embodying the quark-quark interaction was proportional to

$$\lambda^2/k^4, \tag{1}$$

where λ is a constant with the dimensions of mass and k is the gluon four-momentum. As a result the Schwinger mechanism[4] manifestly occurred, and it was shown that any charged particle was totally screened by vacuum polarization effects. In addition, explicit calculations of the deep-inelastic electroproduction structure functions in perturbation theory were in agreement with Bjorken scaling with corrections of $O(q^{-4})$, where q is the virtual photon four-momentum. These features of the Abelian model will also be shown to be true in the non-Abelian version. In addition, we shall argue that the quarks can be nonrelativistic within hadrons and that the spectrum of states has linearly rising Regge trajectories.

In spite of these salutary properties an interaction of the form of Eq. (1) could be questioned because of well-known[5] indefinite-metric difficulties which result in the violation of unitarity. While an optimist may hope that the nonappearance of colored gluons in asymptotic (color singlet) states might eliminate unitarity problems it is almost certain that the approximation techniques which will necessarily be used to find the bound states will lead to the occurrence of negative-metric states. Whether these states are "real" or artifacts of the approximation will not be clear. In view of this we suggested[3] that the gluon propagator be taken in principal value rather than as a Feynman propagator:

$$\mathrm{P}\frac{\lambda^2}{k^4} \equiv \frac{\lambda^2}{2}\left[\frac{1}{(k^2+i\epsilon)^2}+\frac{1}{(k^2-i\epsilon)^2}\right]. \tag{2}$$

As a result unitarity is maintained order by order in perturbation theory. Gluons do not appear in asymptotic states. All components of the vector-gluon propagator are "Coulombized" and the gluon field reduced to the embodiment of a direct quark interaction. There are a number of other decided advantages to principal-value propagators in the present context: (1) no color singlet states composed solely of gluons, (2) the elimination of substantial infrared divergences, (3) the suppression of corrections to Bjorken scaling in the electroproduction structure functions by a factor of q^2

vis-à-vis the corresponding Feynman-propagator result which sets the stage for precocious scaling, and (4) the elimination of closed loops of vector gluons and thus the elimination of Faddeev-Popov ghost loops.

In Sec. II we give a brief recapitulation of the Abelian model. In Sec. III we describe the canonical properties of the non-Abelian model. In Sec. IV we describe the qualitative features of the model and describe an approximation technique which appears to be naturally adopted to "solving" the theory. We shall restrict our discussion to the color binding interaction and defer the introduction of other interactions to a later work. The properties of the bound states in the non-Abelian model are currently under study and will be the subject of the next report.

II. ABELIAN MODEL

The possibility that the physical particle spectrum of a field theory consisted only of neutral states and did not include states of charged fields was first investigated in massless two-dimensional quantum electrodynamics.[6] In that case the absence of the "electron" from the gauge-invariant physical particle spectrum was directly related to the acquisition of a mass by the photon via the Schwinger mechanism. The Schwinger mechanism was manifest in the lowest-order contribution to the vacuum polarization (Fig. 1), and, taking account of the dimensionality of the coupling constant, $e \sim$ mass, could almost be considered a consequence of dimensional analysis. These vacuum polarization effects led to the total screening of the "electronic" charge, and, as a result, the "electron" was removed from the gauge-invariant physical particle spectrum. Our Abelian and non-Abelian models will display a similar pattern of events.

The Lagrangian of the Abelian model contains two gluon fields, $A^1_\mu(x)$ and $A^2_\mu(x)$, and the quark field $\psi(x)$:

$$\mathcal{L} = -\tfrac{1}{2}F^1_{\mu\nu}F^2_{\mu\nu} - \tfrac{1}{2}\lambda^2 A^2_\mu A^2_\mu + \bar\psi(i\nabla\!\!\!/ - gA\!\!\!/^1 - m)\psi\,, \tag{3}$$

where for typographic convenience we denote the inner product of four vectors, $a\cdot b = a_\mu b_\mu = a_0 b_0 - \vec{a}\cdot\vec{b}$ throughout, λ is a constant with the dimensions of mass, g is dimensionless, and $F^i_{\mu\nu} = \partial_\nu A^i_\mu - \partial_\mu A^i_\nu$.

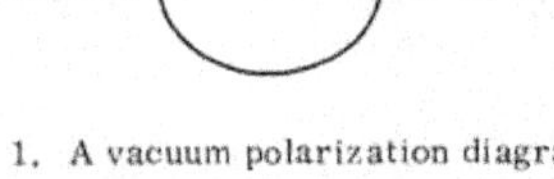

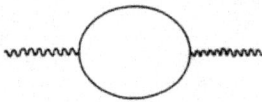

FIG. 1. A vacuum polarization diagram.

Following the canonical procedure we find the equations of motion;

$$\partial_\mu F^1_{\mu\nu} + \lambda^2 A^2_\nu = 0\,, \tag{4}$$

$$\partial_\mu F^2_{\mu\nu} + gJ_\nu = 0\,, \tag{5}$$

$$(i\nabla\!\!\!/ - gA\!\!\!/^1 - m)\psi = 0\,, \tag{6}$$

and nonzero equal-time commutation relations [in the Coulomb gauge $\vec\nabla\cdot\vec{A}^1 = 0$; note $\partial_\mu A^2_\mu = 0$ by Eq. (4)]

$$[F^1_{0i}(x), A^2_j(y)] = i\Delta^{tr}_{ij}(x-y)\,, \tag{7}$$

$$[F^2_{0i}(x), A^1_j(y)] = i\Delta^{tr}_{ij}(x-y)\,, \tag{8}$$

with $i, j = 1, 2, 3$ and

$$\Delta^{tr}_{ij}(x-y) = \int \frac{d^3k}{(2\pi)^3} e^{i\vec{k}\cdot(\vec{x}-\vec{y})}\left(\delta_{ij} - \frac{k_i k_j}{|\vec{k}|^2}\right). \tag{9}$$

It is clear from the equations of motion, Eqs. (4) and (5), tnat A^2_μ may be eliminated to obtain

$$\Box\partial_\mu F^1_{\mu\nu} + g\lambda^2 J_\nu = 0\,. \tag{10}$$

The form of the quark-gluon interaction and Eq. (10) show that only the Green's function of A^1_μ is relevant to quark-quark scattering. The perturbation theory rules of QED may be used if the photon propagator is replaced with the gluon propagator for A^1_μ:

$$iG^{11}_{\mu\nu}(k) = \frac{i\lambda^2(g_{\mu\nu} - \chi k_\mu k_\nu/k^2)}{k^4}\,, \tag{11}$$

where χ is constant, determined by the gauge choice.

In Ref. 3 we showed that choosing $G^{11}_{\mu\nu}$ to be a principal-value propagator allowed us to develop a perturbation theory which was unitary order by order:

$$G^{11}_{\mu\nu}(k^2) = \tfrac{1}{2}[G^{11}_{\mu\nu}(k^2 + i\epsilon) + G^{11}_{\mu\nu}(k^2 - i\epsilon)]\,. \tag{12}$$

In addition, the equivalent of the Nambu representation of a Feynman diagram was given and some features of the perturbation theory discussed. Of particular interest was a calculation of the deep-inelastic electroproduction structure functions which scaled in the Bjorken limit. Leading corrections to scaling were of $O(q^{-4})$ as $q^2 \to \infty$ with q being the virtual photon four-momentum, and were given by the diagrams of Fig. 2(b), 2(c), and 2(d). This is to be contrasted with the logarithmic deviations from scaling found in pseudoscalar or vector meson models previously studied.[7]

The Schwinger mechanism manifestly occurred in low orders of perturbation theory. As a result quarks (and all charged objects) are removed from the gauge-invariant spectrum of physical states. The total screening of charge can be seen from the following argument.[3] Consider a spatially bounded system of charge density ρ. The total charge is

$$Q = \int d^3x\, \rho(x) \tag{13}$$

$$= \frac{-1}{g\lambda^2} \int d^3x\, \Box \nabla^2 A_0^1 \tag{14}$$

using the equations of motion in the Coulomb gauge. By Gauss's law

$$Q = \frac{-1}{g\lambda^2} \int d\vec{S} \cdot \vec{\nabla} \Box A_0^1 . \tag{15}$$

From the definition of a Green's function, we have

$$A_0^1(x) = \int d^4y\, G_{00}^{11}(x-y)\rho(y) \tag{16}$$

in the Coulomb gauge. If, for simplicity, we choose ρ to describe a static point quark charge and use the free gluon propagator [Eq. (11)], then $Q \neq 0$. However, if we take account of the effect of vacuum polarization processes (the Schwinger mechanism) we find $A_0^1(x)$ is a monotonically decreasing function of $|\vec{x}|$ for large $|\vec{x}|$ and consequently $Q = 0$ in the limit where the integration surface is taken to infinity in Eq. (15). Thus the spectrum of physical states does not include states of nonzero charge. In the next section we shall show that the proof of quark confinement is essentially the same in the non-Abelian model.

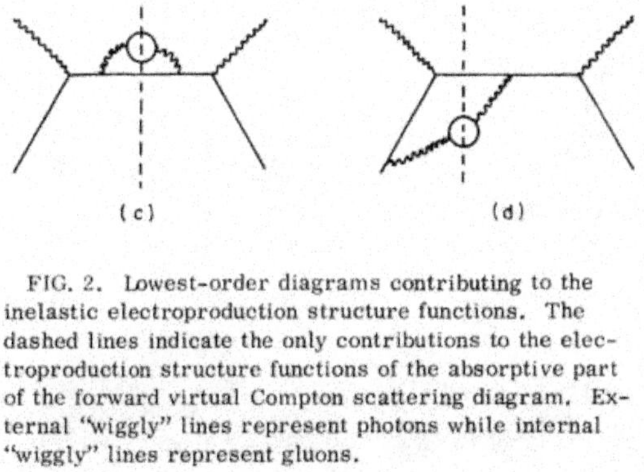

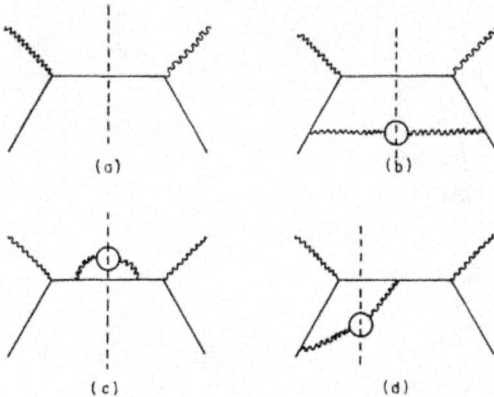

FIG. 2. Lowest-order diagrams contributing to the inelastic electroproduction structure functions. The dashed lines indicate the only contributions to the electroproduction structure functions of the absorptive part of the forward virtual Compton scattering diagram. External "wiggly" lines represent photons while internal "wiggly" lines represent gluons.

III. NON-ABELIAN MODEL

The non-Abelian model for the color sector of hadronic interactions is a direct generalization of the model of the last section.[8] There are two colored Yang-Mills fields, $A_{\mu a}^1(x)$ and $A_{\mu a}^2(x)$, which when regarded as vectors in the adjoint representation of the color group are denoted $\underline{A}_\mu^1$ and $\underline{A}_\mu^2$. The Lagrangian is

$$\begin{aligned}\mathcal{L} &= \tfrac{1}{2}\underline{F}_{\mu\nu}^1 \cdot \underline{F}_{\mu\nu}^2 - \tfrac{1}{2}\underline{F}_{\mu\nu}^2 \cdot (\partial_\mu \underline{A}_\nu^1 - \partial_\nu \underline{A}_\mu^1 + g\underline{A}_\mu^1 \times \underline{A}_\nu^1) \\ &\quad - \tfrac{1}{2}\underline{F}_{\mu\nu}^1 \cdot (\partial_\mu \underline{A}_\nu^2 - \partial_\nu \underline{A}_\mu^2 + g\underline{A}_\mu^1 \times \underline{A}_\nu^2 - g\underline{A}_\nu^1 \times \underline{A}_\mu^2) \\ &\quad - \tfrac{1}{2}\lambda^2 \underline{A}_\mu^2 \cdot \underline{A}_\mu^2 + \bar{\psi}(i\nabla\!\!\!/ + gA\!\!\!/^1 - m)\psi \qquad (17) \\ &= \mathcal{L}_0 + \bar{\psi}(i\nabla\!\!\!/ + gA\!\!\!/^1 - m)\psi , \qquad (18)\end{aligned}$$

with ψ being the quark field.

It is invariant under the local gauge transformation

$$\psi' = S^{-1}\psi , \tag{19}$$

$$A_\mu^{1\prime} = S^{-1}A_\mu^1 S + \frac{i}{g} S^{-1}\partial_\mu S , \tag{20}$$

$$A_\mu^{2\prime} = S^{-1}A_\mu^2 S , \tag{21}$$

$$F_{\mu\nu}^{1\prime} = S^{-1}F_{\mu\nu}^1 S , \tag{22}$$

$$F_{\mu\nu}^{2\prime} = S^{-1}F_{\mu\nu}^2 S , \tag{23}$$

where S is an element in the gauge group G [which is color SU(3) in our case], and A_μ^1 is a matrix in the defining representation of G formed from

$$A_\mu^1 = \underline{A}_\mu^1 \cdot \underline{T} . \tag{24}$$

T_a is a matrix in the defining representation of G satisfying

$$[T_a, T_b] = i f_{abc} T_c , \tag{25}$$

and $\underline{T}$ is a vector formed from such matrices. We note that the homogeneity of the gauge transformation of A_μ^2 allows a mass term to occur in $\mathcal{L}$ without breaking the gauge symmetry. We shall see that the natural gauge-fixing term to add to the Lagrangian has the form

$$-\frac{1}{\beta}\partial_\mu \underline{A}_\mu^1 \cdot \partial_\nu \underline{A}_\nu^2 . \tag{26}$$

The Euler-Lagrange equations of motion are obtained in the canonical manner:

$$(\partial_\mu + g\underline{A}_\mu^1 \times)\underline{F}_{\mu\nu}^1 - \lambda^2 \underline{A}_\nu^2 = 0 , \tag{27}$$

$$(\partial_\mu + g\underline{A}_\mu^1 \times)\underline{F}_{\mu\nu}^2 + g\underline{A}_\mu^2 \times \underline{F}_{\mu\nu}^1 + g\underline{J}_\nu = 0 , \tag{28}$$

$$\underline{F}_{\mu\nu}^1 = \partial_\mu \underline{A}_\nu^1 - \partial_\nu \underline{A}_\mu^1 + g\underline{A}_\mu^1 \times \underline{A}_\nu^1 , \tag{29}$$

$$\underline{F}_{\mu\nu}^2 = \partial_\mu \underline{A}_\nu^2 - \partial_\nu \underline{A}_\mu^2 + g\underline{A}_\mu^1 \times \underline{A}_\nu^2 - g\underline{A}_\nu^1 \times \underline{A}_\mu^2 , \tag{30}$$

$$(i\nabla\!\!\!/ + gA\!\!\!/^1 - m)\psi = 0 . \tag{31}$$

The antisymmetry of $\underline{F}_{\mu\nu}^1$ and $\underline{F}_{\mu\nu}^2$ leads to two conservation laws,

$$\partial_\nu(g\underline{A}^1_\mu\times\underline{F}^1_{\mu\nu}-\lambda^2\underline{A}^2_\nu)=0\,, \tag{32}$$

$$\partial_\nu(\underline{A}^1_\mu\times\underline{F}^2_{\mu\nu}+\underline{A}^2_\mu\times\underline{F}^1_{\mu\nu}+\underline{J}_\nu)=0\,, \tag{33}$$

which can be reexpressed as

$$(\partial_\nu+g\,\underline{A}^1_\nu\times)\underline{A}^2_\nu=0 \tag{34}$$

and

$$(\partial_\nu+g\underline{A}^1_\nu\times)\underline{J}_\nu=0 \tag{35}$$

using the equations of motion. The first of these relations acts in effect as a gauge-fixing term for $\underline{A}^2_\mu$ if a gauge is chosen for $\underline{A}^1_\mu$. The second relation has the familiar form of current-conservation equations in conventional Yang-Mills theories.

We turn now to the derivation of the perturbation-theory rules in the gluon sector. We consider the vacuum-vacuum transition amplitude in the presence of external sources[9]:

$$W(\underline{J}^1_\mu,\underline{J}^2_\mu)=\int\prod_x dA^1_\mu\, dA^2_\nu \exp\left[i\int d^4x\left(\mathfrak{L}_0-\frac{1}{\beta}\partial_\mu\underline{A}^1_\mu\cdot\partial_\nu\underline{A}^2_\nu+\underline{A}^1_\mu\cdot\underline{J}^1_\mu+\underline{A}^2_\mu\cdot\underline{J}^2_\mu\right)\right]. \tag{36}$$

After some functional translations we find

$$W(\underline{J}^1_\mu,\underline{J}^2_\nu)=\exp\left\{-i\int d^4x\,d^4y\left[\underline{J}^1_\mu(x)\cdot G^{12}_{\mu\nu}(x-y)\cdot\underline{J}^2_\nu(y)+\tfrac{1}{2}\underline{J}^1_\mu(x)\cdot G^{11}_{\mu\nu}(x-y)\cdot\underline{J}^1_\nu(y)\right]\right\}, \tag{37}$$

where we have dropped an irrelevant factor independent of $\underline{J}^1_\mu$ and $\underline{J}^2_\mu$ on the right-hand side, and

$$G^{12}_{\mu\nu ab}(x)=-\delta_{ab}\int\frac{d^4k\,e^{-ik\cdot x}}{(2\pi)^4k^2}\left[g_{\mu\nu}+(\beta-1)\frac{k_\mu k_\nu}{k^2}\right] \tag{38}$$

and

$$G^{11}_{\mu\nu ab}(x)=\frac{\lambda^2\delta_{ab}}{(2\pi)^4}\int\frac{d^4k\,e^{-ik\cdot x}}{k^4}\left[g_{\mu\nu}+(\beta^2-1)\frac{k_\mu k_\nu}{k^2}\right], \tag{39}$$

with a and b labeling color indices. The free propagators corresponding to the time-ordered products are

$$\langle TA^1_{\mu a}(x)A^1_{\nu b}(y)\rangle=iG^{11}_{\mu\nu ab}(x-y) \tag{40}$$

and

$$\langle TA^1_{\mu a}(x)A^2_{\nu b}(y)\rangle=iG^{12}_{\mu\nu ab}(x-y)\,. \tag{41}$$

The somewhat unusual Green's functions of Eqs. (40) and (41) have their origin in the canonical equal-time commutation relations which we shall now find.

From Eqs. (27)–(30) we obtain the equations of motion

$$\partial_0\underline{A}^1_k=\underline{F}^1_{0k}+\partial_k\underline{A}^1_0+g\underline{A}^1_k\times\underline{A}^1_0\,, \tag{42}$$

$$\partial_0\underline{A}^2_k=\underline{F}^2_{0k}+\partial_k\underline{A}^2_0+g\underline{A}^2_k\times\underline{A}^1_0-g\underline{A}^2_0\times\underline{A}^1_k\,, \tag{43}$$

$$\partial_0\underline{F}^1_{0k}=(\partial_i+g\underline{A}_i\times)\underline{F}^1_{ik}-g\underline{A}^1_0\times\underline{F}^1_{0k}+\lambda^2\underline{A}^2_k\,, \tag{44}$$

$$\partial_0\underline{F}^2_{0k}=(\partial_i+g\underline{A}^1_i\times)\underline{F}^2_{ik}-g\underline{A}^1_0\times\underline{F}^2_{0k}-g\underline{A}^2_\mu\times\underline{F}^1_{\mu k}-g\underline{J}_k\,, \tag{45}$$

and equations of constraint

$$\underline{F}^1_{ik}=\partial_i\underline{A}^1_k-\partial_k\underline{A}^1_i+g\underline{A}^1_i\times\underline{A}^1_k\,, \tag{46}$$

$$\underline{F}^2_{ik}=\partial_i\underline{A}^2_k-\partial_k\underline{A}^2_i+g\underline{A}^1_i\times\underline{A}^2_k-g\underline{A}^1_k\times\underline{A}^2_i\,, \tag{47}$$

$$(\partial_i+g\underline{A}^1_i\times)\underline{F}^1_{i0}+\lambda^2\underline{A}^2_0=0\,, \tag{48}$$

$$(\partial_i+g\underline{A}^1_i\times)\underline{F}^2_{i0}+g\underline{A}^2_i\times\underline{F}^1_{i0}-g\underline{J}_0=0\,. \tag{49}$$

The Lagrangian indicates that the canonical momenta are

$$\underline{\Pi}^1_j=\underline{F}^2_{0j} \tag{50}$$

and

$$\underline{\Pi}^2_j=\underline{F}^1_{0j}\,, \tag{51}$$

for $j=1,2,3$ with $\underline{\Pi}^i_j$ conjugate to $\underline{A}^i_j$, and $\underline{A}^i_0$ having no conjugate momentum for $i=1,2$. However, the equations of constraint indicate that not all components are independent. We now find the independent components. Let us define

$$\underline{F}^a_{0i}=\underline{F}^{aT}_{0i}+\underline{F}^{aL}_{0i} \tag{52}$$

and

$$\underline{F}^{aL}_{0i}=\partial_i\underline{\varphi}^a\,, \tag{53}$$

where

$$\partial_i\underline{F}^{aT}_{0i}=0\,. \tag{54}$$

Then Eq. (48) gives

$$(\partial_i+g\underline{A}^1_i\times)\partial_i\underline{\varphi}^1-\lambda^2\underline{A}^2_0=-g\underline{A}^1_i\times\underline{F}^{1T}_{i0} \tag{55}$$

and Eq. (49) gives

$$(\partial_i+g\underline{A}^1_i\times)\partial_i\underline{\varphi}^2+g\underline{A}^2_i\times\partial_i\underline{\varphi}^1$$
$$=g\underline{A}^1_i\times\underline{F}^{2T}_{i0}+g\underline{A}^2_i\times\underline{F}^{1T}_{0i}-g\underline{J}_0\,. \tag{56}$$

Rewriting Eqs. (42) and (43) after taking the divergence with respect to spatial components gives

$$(\partial_0+g\underline{A}^1_0\times)\partial_k\underline{A}^1_k=(\partial_k+g\underline{A}^1_k\times)\partial_k\underline{A}^1_0+\partial_k\partial_k\underline{\varphi}^1 \tag{57}$$

and

$$(\partial_0+g\underline{A}^1_0\times)\partial_k\underline{A}^2_k+g\underline{A}^2_0\times\partial_k\underline{A}^1_k$$
$$=\partial_k\partial_k\underline{A}^2_0+g\underline{A}^2_k\times\partial_k\underline{A}^1_0+gA^1_k\times\partial_k\underline{A}^2_0+\partial_k\partial_k\underline{\varphi}^2 \tag{58}$$

If we choose the Coulomb gauge, $\vec{\nabla}\cdot\underline{A}^1=0$, then

$$\partial_k\partial_k\underline{A}_0^1+g\underline{A}_k^1\times\partial_k\underline{A}_0^1+\partial_k\partial_k\underline{\phi}^1=0 \tag{59}$$

and

$$(\partial_0+g\underline{A}_0^1\times)\partial_k\underline{A}_k^2-\partial_k\partial_k\underline{A}_0^2-g\underline{A}_k^2\times\partial_k\underline{A}_0^1-g\underline{A}_k^1\times\partial_k\underline{A}_0^2 -\partial_k\partial_k\underline{\phi}^2=0\,, \tag{60}$$

thus determining $\underline{A}_0^1$ and $\underline{A}_0^2$. Suppose we now define

$$\underline{\vec{A}}^2=\underline{\vec{A}}^{2T}+\underline{\vec{A}}^{2L}\,, \tag{61}$$

$$\underline{\vec{A}}^{2L}=\vec{\nabla}\underline{\phi}^3\,, \tag{62}$$

with

$$\vec{\nabla}\cdot\underline{\vec{A}}^{2T}=0 \tag{63}$$

Taking the divergence of Eq. (44) leads to our final equation for dependent variables

$$\lambda^2\partial_k\partial_k\underline{\phi}^3=\partial_0\partial_k\partial_k\underline{\phi}^1+g\partial_k(\underline{A}_\mu^1\times\underline{F}_{\mu k}^1)\,. \tag{64}$$

The independent dynamical variables are thus seen to be F_{0i}^{1T}, F_{0i}^{2T}, A_i^{1T}, and A_i^{2T}. Their equal-time commutation relations are

$$[F_{0ia}^{1T}(x),A_{jb}^2(y)]=i\delta_{ab}\Delta_{ij}^{\rm tr}(x-y)\,, \tag{65}$$

$$[F_{0ia}^{2T}(x),A_{jb}^1(y)]=i\delta_{ab}\Delta_{ij}^{\rm tr}(x-y)\,, \tag{66}$$

with $i,j=1,2,3$, $\Delta_{ij}^{\rm tr}$ given by Eq. (9), and a and b are color indices. All other commutators of the forms $[A^1,A^1]$, $[A^2,A^2]$, $[F^1,F^1]$, $[F^2,F^2]$, $[F^1,F^2]$ are zero.

We return to our development of perturbation-theory rules. The cubic and quartic gluon vertices of our model are given by (see Fig. 3)

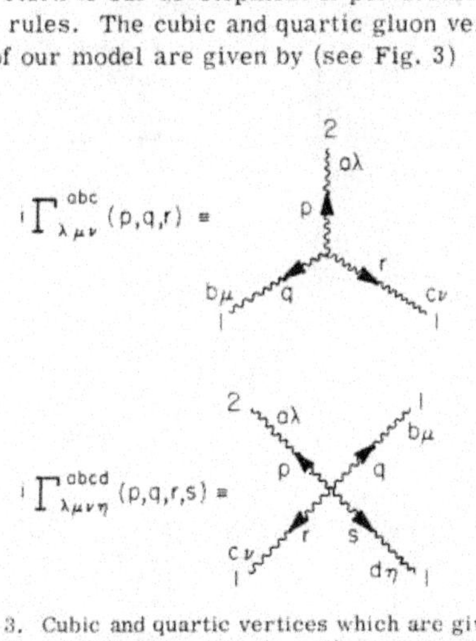

FIG. 3. Cubic and quartic vertices which are given in Eqs. (67) and (68). They introduce $1/r$ potentials in the model and may have an important effect in the baryon spectrum. The numbers 1 and 2 indicate fields $\underline{A}_\mu^1$ and $\underline{A}_\mu^2$, respectively, while p, q, r, and s are momenta, and a, b, c, and d are color indices.

$$i\Gamma^{abc}_{\lambda\mu\nu}(p,q,r)=f^{abc}[g_{\nu\lambda}(r_\mu-p_\mu)+g_{\mu\lambda}(p_\nu-q_\nu) +g_{\mu\nu}(q_\lambda-r_\lambda)]\,, \tag{67}$$

with $p+q+r=0$, and

$$\begin{aligned}i\Gamma^{abcd}_{\lambda\mu\nu\eta}(p,q,r,s)=&-if^{abf}f^{cdf}(g_{\lambda\nu}g_{\nu\eta}-g_{\eta\mu}g_{\lambda\nu})\\&-if^{acf}f^{bdf}(g_{\lambda\eta}g_{\nu\mu}-g_{\nu\lambda}g_{\mu\eta})\\&-if^{adf}f^{bcf}(g_{\lambda\eta}g_{\nu\mu}-g_{\lambda\mu}g_{\nu\eta})\,,\end{aligned} \tag{68}$$

with $p+q+r+s=0$.

The Faddeev-Popov ghost loops will not be relevant to our line of development so we omit their discussion. The necessity for their introduction[10] is closely related to the requirement of unitarity in Yang-Mills theories. In the present model unitarity will be necessarily violated irrespective of the ghost loops if the Green's functions [Eqs. (38) and (39)] pole ambiguities are resolved by using Feynman's $i\epsilon$ procedure. To avoid unitarity violation we have suggested an alternative procedure where the Green's function singularities are taken in principal value,

$$G^{kL}_{\mu\nu ab}(k^2)=\tfrac{1}{2}[G^{kL}_{\mu\nu ab}(k^2+i\epsilon)+G^{kL}_{\mu\nu ab}(k^2-i\epsilon)]\,, \tag{69}$$

in momentum space (cf. the Appendix). This choice has the advantage stated in the Introduction. The effects are the same as in the Abelian model[3] and may be summarized as: (1) Only states composed solely of quarks contribute to unitarity sums, (2) gluons do not appear in asymptotic states, (3) unitarity is achieved but at the price of possible advanced effects whose range is limited to hadronic dimensions and thus apparently unobservable, and (4) nonscaling corrections to Bjorken scaling in the deep-inelastic electroproduction structure functions are suppressed by a factor of q^2 vis-à-vis the corresponding result using Feynman propagators with q being the virtual photon four-momentum.

A novel feature of the use of principal-value propagators in non-Abelian models is the elimination of closed loops composed solely of gluons. If we consider a subdiagram consisting of a gluon loop with p lines, then Eq. (51) of Ref. 3 gives the Feynman parameter representation

$$I=\int_{-\infty}^{\infty}\prod_{j=1}^{p}\alpha_j\,d\alpha_j\,\frac{\epsilon(\alpha_1\alpha_2\cdots\alpha_p C)}{C^2}\,Ne^{iD/C}\,, \tag{70}$$

where C is a polynomial consisting of Feynman parameters only, while D contains scalar products of external momenta, N symbolizes appropriate numerator factors, and $\epsilon(\alpha)=\pm1$ if $\alpha\gtrless0$. Since N can be written as a sum of terms each of which is homogeneous in the Feynman parameters, we can take N to be homogeneous without loss of generality. Then scaling all parameters with u, assuming

$$N(u\alpha_1, u\alpha_2, \ldots, u\alpha_p) = u^r N(\alpha_1, \alpha_2, \ldots, \alpha_p), \quad (71)$$

with r an integer, and using

$$\int_0^\infty \frac{du}{u}\,\delta\left(1 - \frac{|\alpha_1 + \alpha_2 + \cdots + \alpha_p|}{u}\right) = 1 \quad (72)$$

we find

$$I = \Gamma(r + 2p - 2L) \times \int_{-\infty}^{\infty} \frac{\prod_{j=1}^{p} \alpha_j\, d\alpha_j\, \epsilon(\alpha_1\alpha_2\cdots\alpha_p C) N \delta\left(1 - \left|\sum_k \alpha_k\right|\right)}{C^2(-iD/C)^{r+2p-2L}}, \quad (73)$$

with L = number of loops = 1. Suppose we let $\alpha_j \to -\alpha_j$ for all j in I. Then we find $I = -I$ or

$$I = 0. \quad (74)$$

Thus any closed loop containing only principal-value propagators is zero. Since Faddeev-Popov ghosts appear only in closed loops and consistency[11] requires we use principal-value propagators for them if we use such propagators for gluons, we see that ghosts do not appear in our model. Physically we can understand this result if we remember that ghost loops were introduced to cure problems arising from contributions to unitarity sums of "opened" gluon loops.[10] In our model "opened" loops do not contribute to unitarity sums in any case so the raison d'être for ghosts is lacking.

We now derive the Ward-Takahashi-Slavnov identities using functional methods. Since we take our gluon propagators in principal value it might appear that our use of functional techniques is unjustified. We shall take the view that the functional representation of the vacuum-vacuum transition amplitude embodies the combinatorics of perturbation theory and acts as a generating function for identities, such as the Ward-Takahashi-Slavnov identities. Thus, questions of convergence of functional integrals are irrelevant—the important question is whether identities are valid in perturbation theory.

We define $W(J)$, the vacuum-vacuum transition amplitude, by

$$W(J) = \int \prod_x dA_\mu^1\, dA_\mu^2\, d\psi\, d\bar\psi \exp\left(i\int \tilde{\mathcal{L}}\, dx\right), \quad (75)$$

with

$$\tilde{\mathcal{L}} = \mathcal{L} - \frac{1}{\beta}\partial_\mu \underline{A}_\mu^1 \cdot \partial_\nu \underline{A}_\nu^2 + \underline{A}_\mu^1 \cdot \underline{J}_\mu^1 + \underline{A}_\mu^2 \cdot \underline{J}_\mu^2 + \bar\psi\eta + \bar\eta\psi,$$

with $\mathcal{L}$ given by Eq. (17). Under the infinitesimal gauge variation

$$\underline{A}_\mu^1 \to \underline{A}_\mu^1 - (\partial_\mu + g\underline{A}_\mu^1 \times)\underline{\theta}, \quad (76)$$

$$\underline{A}_\mu^2 \to \underline{A}_\mu^2 - g\underline{A}_\mu^2 \times \underline{\theta}, \quad (77)$$

$$\psi \to \psi - ig\theta\psi, \quad (78)$$

$$\bar\psi \to \bar\psi + ig\bar\psi\theta, \quad (79)$$

with $\theta = \underline{T}\cdot\underline{\theta}$, $\mathcal{L}$ is invariant but the remaining terms in $\tilde{\mathcal{L}}$ lead to

$$\delta\tilde{\mathcal{L}} = \frac{1}{\beta}[(\partial_\mu + g\underline{A}_\mu^1 \times)\partial_\nu\partial_\mu \underline{A}_\nu^2 + g\underline{A}_\nu^2 \times \partial_\mu\partial_\nu \underline{A}_\mu^1]\cdot\underline{\theta} - (\partial_\mu + g\underline{A}_\mu^1\times)\underline{J}_\mu^1\cdot\underline{\theta} - g\underline{J}_\mu^2 \times \underline{A}_\mu^2\cdot\underline{\theta} + ig\bar\psi\theta\eta - ig\bar\eta\theta\psi. \quad (80)$$

Since a transformation of the integration variables does not change the value of the functional integral, the variation of W with respect to θ can be taken to be zero and our equivalent of the Ward-Takahashi-Slavnov identity is

$$\left\{\frac{1}{\beta}\left[D_\nu\left(\frac{\delta}{i\delta \underline{J}_\alpha^1}\right)\partial_\nu\partial_\mu \frac{\delta}{i\delta \underline{J}_\mu^2} + g\frac{\delta}{i\delta \underline{J}_\nu^2}\times\partial_\nu\partial_\mu\frac{\delta}{i\delta \underline{J}_\mu^1}\right] + D_\mu\left(\frac{\delta}{i\delta \underline{J}_\alpha^1}\right)\underline{J}_\mu^1 - g\underline{J}_\mu^2 \times \frac{\delta}{i\delta \underline{J}_\mu^2} + g\underline{T}\eta\frac{\delta}{\delta\eta} - g\bar\eta\underline{T}\frac{\delta}{\delta\bar\eta}\right\}W = 0, \quad (81)$$

with

$$D_\mu\left(\frac{\delta}{i\delta \underline{J}_\alpha^1}\right) = \partial_\mu + g\frac{\delta}{i\delta \underline{J}_\mu^1}\times. \quad (82)$$

In order to investigate the structure of the gluon propagators we shall obtain the proper vertex identity equivalent to Eq. (81). We focus on the novelties of the gluon sector and neglect the quark field terms in $\tilde{\mathcal{L}}$ and Eq. (81). Let us define

$$W(J) = e^{iZ(J)}, \quad (83)$$

$$\underline{B}_\mu^i = -\frac{\delta Z(J)}{\delta \underline{J}_\mu^i}, \quad i = 1, 2 \quad (84)$$

$$\Gamma(B) = Z(J) + \int d^4x(\underline{J}_\mu^1\cdot\underline{B}_\mu^1 + \underline{J}_\mu^2\cdot\underline{B}_\mu^2), \quad (85)$$

where $\Gamma(B)$ is the generating functional of proper vertices. An immediate consequence is

$$\underline{J}^i_\mu = \frac{\delta\Gamma}{\delta \underline{B}^i_\mu}, \quad i=1,2 \tag{86}$$

and as a result Eq. (81) can be rewritten in the form

$$\frac{1}{\beta}\left[\Box\partial_\mu \underline{B}^2_\mu - g\underline{B}^1_\nu\times\partial_\nu\partial_\mu \underline{B}^2_\mu - g\underline{B}^2_\nu\times\partial_\mu\partial_\nu \underline{B}^1_\mu + g\frac{\delta}{i\delta \underline{J}^1_\nu}\times\partial_\nu\partial_\mu \underline{B}^2_\mu + g\frac{\delta}{i\delta \underline{J}^2_\nu}\times\partial_\nu\partial_\mu \underline{B}^1_\mu\right] - \partial_\mu\frac{\delta\Gamma}{\delta \underline{B}^1_\mu} + \underline{B}^1_\mu\times\frac{\delta\Gamma}{\delta \underline{B}^1_\mu} + \underline{B}^2_\mu\times\frac{\delta\Gamma}{\delta \underline{B}^2_\mu} = 0. \tag{87}$$

If we apply $\delta/\delta\underline{B}^1_\alpha$ to Eq. (87) and set $\underline{B}^i_\mu = 0$ afterwards, we find

$$-\partial_\mu \frac{\delta^2\Gamma}{\delta \underline{B}^1_\alpha \delta \underline{B}^1_\mu}\bigg|_{B^1=B^2=0} = 0. \tag{88}$$

The second-order functional derivative of Γ is the inverse of the full propagator $G^{11\prime}_{\mu\nu ab}$ and Eq. (88) implies that the proper part of $(G^{11\prime}_{\mu\nu ab})^{-1}$ is purely transverse. We note that the "free" propagator (Eq. 39) contribution to $(G^{11\prime}_{\mu\nu ab})^{-1}$ is not one-particle irreducible and thus not constrained by Eq. (88). Therefore we find the general form

$$G^{11\prime}_{\mu\nu ab}(k) = \delta_{ab}\left(g_{\mu\nu} - \frac{k_\mu k_\nu}{k^2}\right)G^{11}(k^2) + \delta_{ab}\beta^2\lambda^2\frac{k_\mu k_\nu}{k^6}, \tag{89}$$

so that the longitudinal part of the full propagator is not renormalized.

The longitudinal part of the full propagator $G^{12\prime}_{\mu\nu ab}(k)$ is also not renormalized. This may be seen by applying $\delta/\delta\underline{B}^2_\mu$ to Eq. (87) and setting $\underline{B}^i_\mu = 0$ afterwards:

$$\frac{1}{\beta}\Box\partial_\mu\delta^4(x-y) - \partial_\nu\frac{\delta^2\Gamma}{\delta \underline{B}^2_\mu \delta \underline{B}^1_\nu}\bigg|_{B^1=B^2=0} = 0. \tag{90}$$

This implies

$$(G^{12\prime}_{\mu\nu ab})^{-1} = \frac{\delta_{ab}(g_{\mu\nu} - k_\mu k_\nu/k^2)}{G^{12}} - \frac{k_\mu k_\nu \delta_{ab}}{\beta} \tag{91}$$

or

$$G^{12\prime}_{\mu\nu ab}(k) = \delta_{ab}\left(g_{\mu\nu} - \frac{k_\mu k_\nu}{k^2}\right)G^{12}(k) - \beta\delta_{ab}\frac{k_\mu k_\nu}{k^4}. \tag{92}$$

Having now developed the general form of the propagators we now will define the gluon vacuum polarization tensors,

$$\Pi^{11}_{\mu\nu ab}(k) = [G^{11\prime}_{\mu\nu ab}(k)]^{-1} - [G^{11}_{\mu\nu ab}(k)]^{-1}, \tag{93}$$

$$\Pi^{12}_{\mu\nu ab}(k) = [G^{12\prime}_{\mu\nu ab}(k)]^{-1} - [G^{12}_{\mu\nu ab}(k)]^{-1}, \tag{94}$$

which are transverse by our previous discussion:

$$k_\mu \Pi^{11}_{\mu\nu ab} = k_\mu \Pi^{12}_{\mu\nu ab} = 0. \tag{95}$$

Rather than write the Schwinger-Dyson equations for our polarization tensors we have given a diagrammatic representation in Fig. 4.

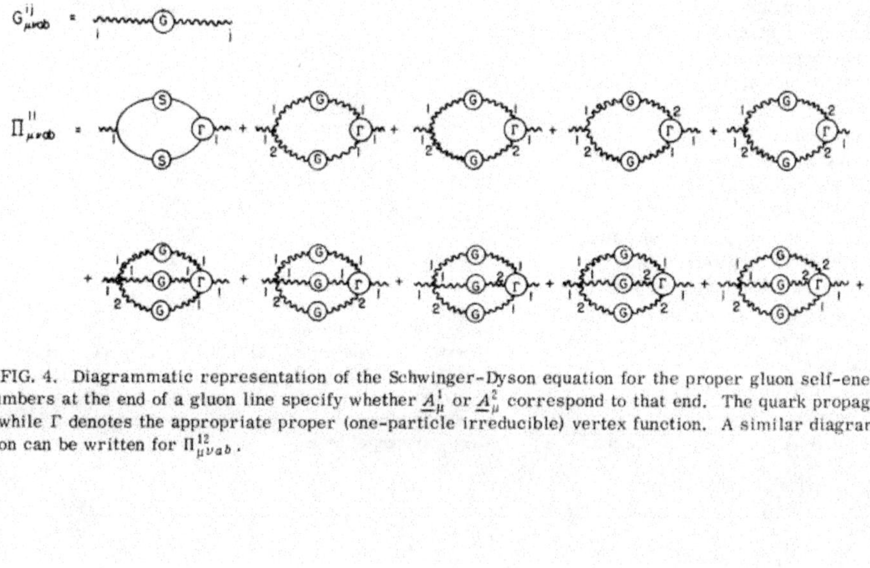

FIG. 4. Diagrammatic representation of the Schwinger-Dyson equation for the proper gluon self-energy, $\Pi^{11}_{\mu\nu ab}$. The numbers at the end of a gluon line specify whether $\underline{A}^1_\mu$ or $\underline{A}^2_\mu$ correspond to that end. The quark propagator is denoted S while Γ denotes the appropriate proper (one-particle irreducible) vertex function. A similar diagrammatic expression can be written for $\Pi^{12}_{\mu\nu ab}$.

IV. OBSERVATIONS

The Schwinger mechanism forces quark confinement to bound color singlet states in a manner which is identical to the Abelian case as described in Sec. II. In order to demonstrate that only color singlets exist in the gauge-invariant physical particle spectrum it is sufficient to show

$$\underline{Q}\psi_{\text{phys}} = 0\,, \tag{96}$$

where

$$\underline{Q} = \int d^3x\, \underline{J}_0(x) \tag{97}$$

for any physical state ψ_{phys} corresponding to a spatially localized distribution of quarks. We consider a single static quark located at the origin and choose to work in the Coulomb gauge ($\vec{\nabla}\cdot\vec{A}^1 = 0$). Then the time components of the equations of motion [Eqs. (27) and (28)] lead to (at large distance)

$$\Box\nabla^2 \underline{A}_0^1 = g\lambda^2 \underline{J}_0 \tag{98}$$

if we take into account the elimination of gluons' degrees of freedom through the choice of principal-value propagators and their consequent inability to act as sources. We may now repeat the arguments of Eqs. (13)–(16) for the Abelian case after noticing the occurrence of the Schwinger mechanism in the non-Abelian case which can be verified in low orders of perturbation theory for $\Pi^{11}_{\mu\nu ab}$. Thus the expectation value of the charge in the one-quark state is zero. Since the one-quark state is a charge eigenstate, we find Eq. (96) to be true in this case and more generally through the additivity of the charge operator. Thus only color singlet bound states of quarks are physical.[12]

While the infrared behavior of the theory leads to quark confinement, the ultraviolet behavior allows the quarks to appear quasifree. This is particularly noticeable when we take $\lambda^2 = 0$ in our Lagrangian and examine the corresponding perturbation theory. Taking $\lambda^2 = 0$ is equivalent to examining the short-distance behavior of the theory. The only diagrams which exist in this limit are given in Fig. 5. The quark sector of the theory is free. The only nontree structures are one-quark-loop diagrams for the scattering of gluons associated with A_μ^2 (which of course can only be generated by a hypothetical external source). (As a point of comparison we have shown in Fig. 6 the additional diagrams which would occur in the even that Feynman propagators were used—these diagrams necessarily involve gluon loops which principal-value propagators force to be zero.) The vital role of the $\lambda^2 A_\mu^2 A_\mu^2$ term in the Lagrangian in generating the interacting theory and the fact that λ^2 has the dimensions of $(\text{mass})^2$ allow a natural approximation procedure in this model. This is perhaps best seen within the context of deep-inelastic electroproduction. Just as in the Abelian case we find that the structure functions scale with leading corrections of $O(q^{-4})$, where q equals the virtual photon four-momentum. We can establish a parton picture of scattering wherein the photon is absorbed on one of the quasifree nucleon constituents [as in Fig. 2(a)] if $|q^2| \gg g^2\lambda^2$. Then leading corrections to such a picture [e.g., the diagrams of Fig. 2(b)–2(d)] will be suppressed by $(g^2\lambda^2/q^2)^2$. Thus the dimensional nature of the effective coupling constant allows a particularly simple picture to exist of the region of large spacelike virtual photon mass and the parton picture emerges as a natural approximation.

The k^{-4} form of the quark interaction also appears to have decidedly good features as far as the bound-state structure is concerned. Ignoring the numerator tensor (which does not affect our conclusions), we find the Fourier transform of the gluon propagator,

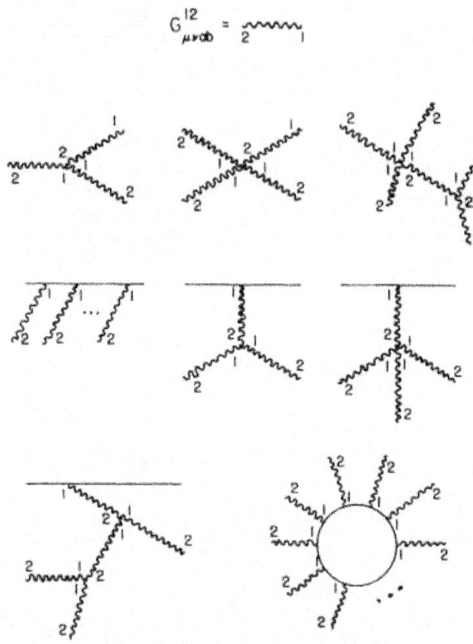

FIG. 5. Some examples of the surviving diagrams in the $\lambda^2 = 0$ limit of the non-Abelian model with principal-value gluon propagators. Except for the class of one-fermion-loop diagrams only tree diagrams exist in this limit. Note that there are no four (or more) external quark line diagrams and no two (or more) external $\underline{A}_\mu^1$ gluon "external" lines.

$$G(k) = \mathrm{P}\frac{1}{k^4}, \tag{99}$$

to be proportional to

$$\tilde{G}(x) = \theta(x^2). \tag{100}$$

Since $\tilde{G}$ has a smooth finite limit as $x^2 \to 0$, the short-distance limit, arguments can be made[13] that low-mass bound states can occur in this model. In addition, Dalitz[14] has pointed out that the linearity of trajectories on the Chew-Frautschi plot would follow from a flat-bottomed, smooth interaction—a criterion which is met by Eq. (100). [It is interesting to note that had we used a Feynman propagator rather than principal value, then $\tilde{G}$ would have been $\ln(x^2)$ and thus the general criterion just stated would not have been met. This would appear to be another point in favor of our choice of principal-value propagators.]

Another property which is desirable in the bound-state solutions is nonrelativistic motion of the bound-state constituents.[15] Again an interaction of the form of Eq. (99) appears to realize this feature—even in the strong-binding limit. To see this we shall first take account of the Schwinger mechanism and in the spirit of Hartree-Fock theory modify the quark interaction to

$$G'(k) = \mathrm{P}\frac{1}{(k^2 - \mu^2)^2}. \tag{101}$$

If we now take Eq. (101) to be the Green's function for the effective gluon field and calculate the "Coulomb potential" of a static, point quark source located at the origin we find

$$\varphi(r) = \frac{\varphi_0}{\mu} e^{-\mu r}, \tag{102}$$

where ϕ_0 is a constant independent of μ. In the limit $\mu \to 0$ we find

$$\phi(r) \cong \varphi_0\left(\frac{1}{\mu} - r + \cdots\right). \tag{103}$$

The first two terms of Eq. (103) correspond to choosing Eq. (99) rather than Eq. (101) as the gluon Green's function (in the limit $\mu \to 0$). Equation (102) includes vacuum polarization effects which damp the interaction at large distances. Thus Eq. (102) imperfectly reflects the possibility that a quark-antiquark pair can separate and induce another quark-antiquark pair to be created from the vacuum so that two color singlet mesons will result (presuming it is energetically favored). At shorter distances Eq. (102) appears to be a reasonable approximation. This exponential potential was studied within the framework of the Schrödinger equation in the strong-binding limit (ϕ_0/μ large) by Greenberg.[16] He showed that the average momentum of the bound constituent in the s state satisfied

$$\frac{p}{m} \sim \left(\frac{\mu}{m}\right)^{1/3} \tag{104}$$

with m being the quark mass. Thus for μ/m small the quark motion is self-consistently nonrelativistic.

In conclusion, we have shown that a four-dimensional, Lorentz-invariant second-quantized field theory of hadron binding is possible with scaling electroproduction structure functions, only zero-triality physical particle states, and, apparently, linearly rising Regge trajectories and nonrelativistic constituents. A detailed study of the bound states is now in progress.

ACKNOWLEDGMENT

I am grateful to the members of the Newman Laboratory for interesting conversations.

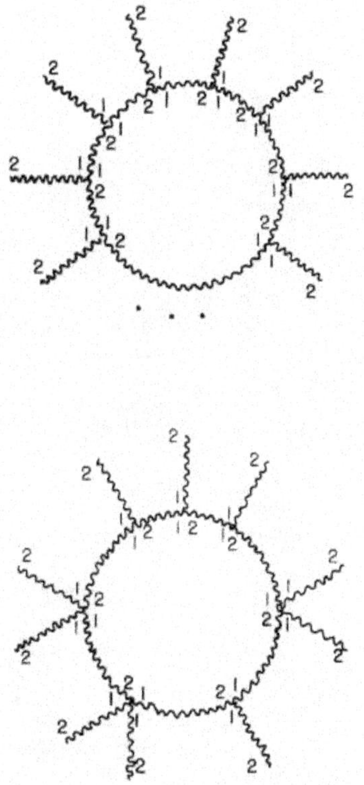

FIG. 6. Some additional diagrams which occur in the $\lambda^2 = 0$ limit of the non-Abelian model if Feynman gluon propagators are used. In addition, there will be Faddeev-Popov ghost-loop diagrams depending on the choice of gauge.

APPENDIX

In Ref. 3 semiclassical arguments based on Dirac's theory of constraints were given to introduce the use of principal-value propagators. We will now describe a second-quantized realization of those arguments for the case of a scalar Klein-Gordon field $\phi(x)$ with the Lagrangian

$$\mathcal{L}=\tfrac{1}{2}(\partial_\mu\phi)^2-\tfrac{1}{2}m^2\phi^2. \tag{A1}$$

The generalization to vector gluons is immediate. The canonical equal-time commutation relations are

$$[\phi,\phi]=[\dot\phi,\dot\phi]=0, \tag{A2}$$

$$[\dot\phi(\vec{x},t),\phi(\vec{y},t)]=-i\delta^3(\vec{x}-\vec{y}). \tag{A3}$$

If we expand $\phi(x)$ in plane waves,

$$\phi(\vec{x},t)=\sum_{\vec{k}}(A_{\vec{k}}^{-}e^{-ik\cdot x}+A_{\vec{k}}^{+}e^{ik\cdot x}), \tag{A4}$$

then the q-number Fourier components $A_{\vec{k}}^{-}$ must satisfy

$$[A_{\vec{k}}^{-},A_{\vec{k}'}^{-}]=[A_{\vec{k}}^{+},A_{\vec{k}}^{+}]=0, \tag{A5}$$

$$[A_{\vec{k}}^{-},A_{\vec{k}'}^{+}]=\delta^3(\vec{k}'-\vec{k}) \tag{A6}$$

for consistency with Eqs. (A2) and (A3). Now the time-ordered product satisfies

$$T(\phi(x)\phi(y))=\epsilon(x_0-y_0)[\phi(x),\phi(y)]+\{\phi(x),\phi(y)\}, \tag{A7}$$

with $\epsilon(x_0)=\pm1$ for $x_0\gtrless0$ and $\{A,B\}=AB+BA$. The first term on the right-hand side is a c number completely determined by Eqs. (A5) and (A6). If the second q-number expression were zero, then we would obtain a principal-value propagator from Eq. (A7):

$$T(\phi(x)\phi(y))=i\int\frac{d^4k}{(2\pi)^4}e^{-ik\cdot(x-y)}\,\mathrm{P}\frac{1}{(k^2-m^2)}. \tag{A8}$$

We therefore require

$$\{\phi(x),\phi(y)\}=0, \tag{A9}$$

with the consequence

$$\{A_{\vec{k}}^{-},A_{\vec{k}'}^{-}\}=\{A_{\vec{k}}^{+},A_{\vec{k}'}^{+}\}=\{A_{\vec{k}}^{-},A_{\vec{k}'}^{+}\}=0. \tag{A10}$$

Equations (A5), (A6), and (A10) imply

$$A_{\vec{k}}^{-}A_{\vec{k}'}^{-}=A_{\vec{k}}^{+}A_{\vec{k}'}^{+}=0, \tag{A11}$$

$$A_{\vec{k}}^{-}A_{\vec{k}'}^{+}=\tfrac{1}{2}\delta^3(\vec{k}-\vec{k}'), \tag{A12}$$

$$A_{\vec{k}}^{+}A_{\vec{k}'}^{-}=-\tfrac{1}{2}\delta^3(\vec{k}-\vec{k}') \tag{A13}$$

for all $\vec{k}$ and $\vec{k}'$. Thus quadratic terms in A and A^+ are reduced to c numbers. It should further be noted that the multiplication rule is not associative.[17] In fact, the multiplication rules of the A and A^+ operators in Eqs. (A11)–(A13) are realized by taking multiplication to be

$$UV=\tfrac{1}{2}[U,V] \tag{A14}$$

for U, V being any $A_{\vec{k}}^{-}$ or $A_{\vec{k}'}^{+}$. If we take an analogy to Lie-algebra theory seriously, where the adjoint representation of an algebra has a multiplication rule defined by commutators

$$\tilde{U}*\tilde{V}=[\tilde{U},\tilde{V}] \tag{A15}$$

then we could call Eqs. (A11)–(A13) the adjoint representation of the Fourier components of ϕ.

The c-number nature of AA, A^+A^+, or AA^+ can be understood physically in the following manner. Since the ϕ field has principal-value propagators it is not associated with a particle but is merely the embodiment of an interaction between other objects (which we have suppressed in our Lagrangian). Consequently an emission of a ϕ field quantum must be directly correlated with a subsequent absorption—it cannot propagate into empty space. The c-number nature of AA^+ reflects this correlation between emission and absorption.

Finally, it should be noted that the existence of a vacuum is inconsistent with Eqs. (A11)–(A13).

---

*Work supported in part by the National Science Foundation.

[1]K. Johnson, Phys. Rev. D 6, 1101 (1972); C. M. Bender, J. E. Mandula, and G. S. Guralnik, Phys. Rev. Lett. 32, 1467 (1974); A. Chodos *et al.*, Phys. Rev. D 9, 3471 (1974); W. A. Bardeen *et al.*, *ibid.* 11, 1094 (1975); M. Creutz, *ibid.* 10, 1749 (1974); P. Vinciarelli, Nuovo Cimento Lett. 4, 905 (1972); R. Dashen, B. Hasslacher, and A. Neveu, Phys. Rev. D 10, 4114 (1974); 10, 4130 (1974); 10, 4138 (1974).

[2]Y. Nambu, in *Preludes in Theoretical Physics*, edited by A. de-Shalit, H. Feshbach, and L. Van Hove (North-Holland, Amsterdam, 1966), p. 133; H. J. Lipkin, Phys. Lett. 45B, 267 (1973).

[3]S. Blaha, Phys. Rev. D 10, 4268 (1974).

[4]J. Schwinger, Phys. Rev. 128, 2425 (1962).

[5]A. Pais and G. Uhlenbeck, Phys. Rev. 79, 145 (1950); J. Kiskis, Phys. Rev. D 11, 2178 (1975).

[6]A. Casher, J. Kogut, and L. Susskind, Phys. Rev. D 10, 732 (1974); J. Lowenstein and J. Swieca, Ann. Phys. (N.Y.) 68, 172 (1971).

[7]R. Jackiw and G. Preparata, Phys. Rev. Lett. 22, 975 (1969); S. Adler and W. Tung, *ibid.* 22, 978 (1969); S. Blaha, Phys. Rev. D 3, 510 (1971).

[8]The Lagrangian of Eq. (17) was first written by D. Sinclair as a generalization of the Abelian model of Ref. 3. An alternative non-Abelian model for quark confinement has been suggested by S. K. Kauffmann [Nucl. Phys. B87, 133 (1975)]. I am grateful to Dr. Kauffmann for sending me a copy of his paper prior to publication.

[9]E. Abers and B. W. Lee [Phys. Rep. 9C, 1 (1973)] provide a useful review of conventional Yang-Mills theories.

[10] R. P. Feynman, Acta Phys. Pol. 24, 697 (1963).

[11]B. W. Lee and J. Zinn-Justin [Phys. Rev. D 5, 3121 (1972)] point out that the $i\epsilon$ prescription in their Eq. (2.8) for the ghost loop is dictated by unitarity considerations.

[12]This does not preclude color-singlet states of the gluons from playing a role in the theory. They are not particles but can be exchanged between color-singlet quark states in scattering events. On naive dimensional grounds they should be most important in forward scattering. This leads to the possibility that the Pomeron might possibly be interpreted as a "two-gluon bound state". In the case of wide-angle scattering the predominant mechanism for large momentum transfer would appear to be constituent interchange due to the strong damping effects of k^{-4} propagators on momentum transfer.

[13]M. Böhm, H. Joos, and M. Krammer, in *Recent Developments in Mathematical Physics*, proceedings of the XII Schladming Conference (Acta Phys. Austriaca Suppl. XI), edited by P. Urban (Springer, New York, 1970), p. 3.

[14]R. H. Dalitz, a paper presented at the Topical Conference on Meson Spectroscopy, Philadelphia, 1968 (unpublished).

[15]H. J. Lipkin, Phys. Rep. 8C, 175 (1973).

[16]O. W. Greenberg, Phys. Rev. 147, 1077 (1966).

[17]Nonassociative field operators have been previously used by M. Günaydin and F. Gürsey, Phys. Rev. D 9, 3387 (1974).

REFERENCES

Bjorken, J. D., Drell, S. D., 1964, *Relativistic Quantum Mechanics* (McGraw-Hill, New York, 1965).

Bjorken, J. D., Drell, S. D., 1965, *Relativistic Quantum Fields* (McGraw-Hill, New York, 1965).

Blaha, S., 1998, *Cosmos and Consciousness* (Pingree-Hill Publishing, Auburn, NH, 1998).

_______, 2002, *A Finite Unified Quantum Field Theory of the Elementary Particle Standard Model and Quantum Gravity Based on New Quantum Dimensions™ & a New Paradigm in the Calculus of Variations* (Pingree-Hill Publishing, Auburn, NH, 2002).

_______, 2003, *A Finite Unified Quantum Field Theory of the Elementary Particle Standard Model and Quantum Gravity Based on New Quantum Dimensions™ and a New Paradigm in the Calculus of Variations* (Pingree-Hill Publishing, Auburn, NH, 2003).

_______, 2004, *Quantum Big Bang Cosmology: Complex Space-time General Relativity, Quantum Coordinates™ Dodecahedral Universe, Inflation, and New Spin 0, ½, 1 & 2 Tachyons & Imagyons* (Pingree-Hill Publishing, Auburn, NH, 2004).

______, *2005a, Quantum Theory of the Third Kind: A New Type of Divergence-free Quantum Field Theory Supporting a Unified Standard Model of Elementary Particles and Quantum Gravity based on a New Method in the Calculus of Variations* (Pingree-Hill Publishing, Auburn, NH, 2005).

______, 2005b, *The Metatheory of Physics Theories, and the Theory of Everything as a Quantum Computer Language* (Pingree-Hill Publishing, Auburn, NH, 2005).

______, 2005c, *The Equivalence of Elementary Particle Theories and Computer Languages: Quantum Computers, Turing Machines, Standard Model, Superstring Theory, and a Proof that Gödel's Theorem Implies Nature Must Be Quantum* (Pingree-Hill Publishing, Auburn, NH, 2005).

______, 2006a, *The Foundation of the Forces of Nature* (Pingree-Hill Publishing, Auburn, NH, 2006).

______, 2006b, *A Derivation of ElectroWeak Theory based on an Extension of Special Relativity; Black Hole Tachyons; & Tachyons of Any Spin.* (Pingree-Hill Publishing, Auburn, NH, 2006).

______, 2007a, *Physics Beyond the Light Barrier: The Source of Parity Violation, Tachyons, and A Derivation of Standard Model Features* (Pingree-Hill Publishing, Auburn, NH, 2007).

______, 2007b, *The Origin of the Standard Model: The Genesis of Four Quark and Lepton Species, Parity Violation, the ElectroWeak Sector, Color SU(3), Three Visible Generations of Fermions, and One Generation of Dark Matter with Dark Energy* (Pingree-Hill Publishing, Auburn, NH, 2007).

______, *2008a, A Direct Derivation of the Form of the Standard Model From GL(16) (Pingree-Hill Publishing, Auburn, NH, 2008).*

______, 2008b, *A Complete Derivation of the Form of the Standard Model With a New Method to Generate Particle Masses Second Edition* (Pingree-Hill Publishing, Auburn, NH, 2008)

______, 2009, *The Algebra of Thought & Reality: The Mathematical Basis for Plato's Theory of Ideas, and Reality Extended to Include A Priori Observers and Space-Time Second Edition* (Pingree-Hill Publishing, Auburn, NH, 2009).

______, 2010a, *Operator Metaphysics: A New Metaphysics Based on a New Operator Logic and a New Quantum Operator Logic that Lead to a Mathematical Basis for Plato's Theory of Ideas and Reality* (Pingree-Hill Publishing, Auburn, NH, 2010).

______, 2010b, *The Standard Model's Form Derived from Operator Logic, Superluminal Transformations and GL(16)* (Pingree-Hill Publishing, Auburn, NH, 2010).

______, 2011a, *21st Century Natural Philosophy Of Ultimate Physical Reality* (McMann-Fisher Publishing, Auburn, NH, 2011).

______, 2011b, *All the Universe! Faster Than Light Tachyon Quark Starships & Particle Accelerators with the LHC as a Prototype Starship Drive Scientific Edition* (Pingree-Hill Publishing, Auburn, NH, 2011).

______, 2011c, *From Asynchronous Logic to The Standard Model to Superflight to the Stars* (Blaha Research, Auburn, NH, 2011).

______, 2012a, *From Asynchronous Logic to The Standard Model to Superflight to the Stars volume 2: Superluminal CP and CPT, U(4) Complex General Relativity and The Standard Model, Complex Vierbein General Relativity, Kinetic Theory, Thermodynamics* (Blaha Research, Auburn, NH, 2012).

______, 2012b, *Standard Model Symmetries, And Four And Sixteen Dimension Complex Relativity; The Origin Of Higgs Mass Terms* (Blaha Reasearch, Auburn, NH, 2012).

______, 2013a, *Multi-Stage Space Guns, Micro-Pulse Nuclear Rockets, and Faster-Than-Light Quark-Gluon Ion Drive Starships* (Blaha Research, Auburn, NH, 2013).

______, 2013b, *The Bridge to Dark Matter; A New Sister Universe; Dark Energy; Inflatons; Quantum Big Bang; Superluminal Physics; An Extended Standard Model Based on Geometry* (Blaha Reasearch, Auburn, NH, 2013).

______, 2014a, *Universes and Multiverses: From a New Standard Model to a Physical Multiverse; The Big Bang; Our Sister Universe's Wormhole; Origin of the Cosmological Constant, Spatial Asymmetry of the Universe, and its Web of Galaxies; A Baryonic Field between Universes and Particles; Flatverse Extended Wheeler-DeWitt Equation* (Blaha Reasearch, Auburn, NH, 2014).

______, 2014b, *All the Multiverse! Starships Exploring the Endless Universes of the Cosmos Using the Baryonic Force* (Blaha Research, Auburn, NH, 2014).

______, 2014c, *All the Multiverse! II Between Multiverse Universes: Quantum Entanglement Explained by the Multiverse Coherent Baryonic Radiation Devices – PHASERs Neutron Star Multiverse Slingshot Dynamics Spiritual and UFO Events, and the Multiverse Microscopic Entry into the Multiverse* (Blaha Research, Auburn, NH, 2014).

______, 2015a, *PHYSICS IS LOGIC PAINTED ON THE VOID: Origin of Bare Masses and The Standard Model in Logic, U(4) Origin of the Generations, Normal and Dark Baryonic Forces, Dark Matter, Dark Energy, The Big Bang, Complex General Relativity, A Megaverse of Universe Particles* (Blaha Research, Auburn, NH, 2015).

______, 2015b, *PHYSICS IS LOGIC Part II: The Theory of Everything, The Megaverse Theory of Everything, U(4)⊗U(4) Grand Unified Theory (GUT), Inertial Mass = Gravitational Mass, Unified Extended Standard Model and a New Complex General Relativity with Higgs Particles, Generation Group Higgs Particles* (Blaha Research, Auburn, NH, 2015).

______, 2015c, *The Origin of Higgs ("God") Particles and the Higgs Mechanism: Physics is Logic III, Beyond Higgs – A Revamped Theory With a Local Arrow of Time, The Theory of Everything Enhanced, Why Inertial Frames are Special, Universes of the Mind* (Blaha Research, Auburn, NH, 2015).

______, 2015d, *The Origin of the Eight Coupling Constants of The Theory of Everything: U(8) Grand Unified Theory of Everything (GUTE), S^8 Coupling Constant Symmetry, Space-Time*

Dependent Coupling Constants, Big Bang Vacuum Coupling Constants, Physics is Logic IV (Blaha Research, Auburn, NH, 2015).

______, 2016a, *New Types of Dark Matter, Big Bang Equipartition, and A New U(4) Symmetry in the Theory of Everything: Equipartition Principle for Fermions, Matter is 83.33% Dark, Penetrating the Veil of the Big Bang, Explicit QFT Quark Confinement and Charmonium, Physics is Logic V* (Blaha Research, Auburn, NH, 2016).

______, 2016b, *The Periodic Table of the 192 Quarks and Leptons in The Theory of Everything: The U(4) Layer Group, Physics is Logic VI* (Blaha Research, Auburn, NH, 2016).

______, 2016c, *New Boson Quantum Field Theory, Dark Matter Dynamics, Dark Matter Fermion Layer Mixing, Genesis of Higgs Particles, New Layer Higgs Masses, Higgs Coupling Constants, Non-Abelian Higgs Gauge Fields, Physics is Logic VII* (Blaha Research, Auburn, NH, 2016).

______, 2016d, *Unification of the Strong Interactions and Gravitation: Quark Confinement Linked to Modified Short-Distance Gravity; Physics is Logic VIII* (Blaha Research, Auburn, NH, 2016).

______, 2016e, *MOND Unification of the Strong Interactions and Gravitation II: Quark Confinement Linked to Large-Scale Gravity; Physics is Logic IX* (Blaha Research, Auburn, NH, 2016).

Chrystal, G., 1961, *Textbook of Algebra Part One* (Dover Publications, Inc., New York, 1961).

Eddington, A. S., 1952, *The Mathematical Theory of Relativity* (Cambridge University Press, Cambridge, U.K., 1952).

Fant, Karl M., 2005, *Logically Determined Design: Clockless System Design With NULL Convention Logic* (John Wiley and Sons, Hoboken, NJ, 2005).

Gutzwiller, M. C., 1990, *Chaos in Classical and Quantum Mechanics* (Springer-Verlag, New York, 1990).

Heitler, W., 1954, *The Quantum Theory of Radiation* (Claendon Press, Oxford, UK, 1954).

Huang, Kerson, 1992, *Quarks, Leptons & Gauge Fields 2nd Edition* (World Scientific Publishing Company, Singapore, 1992).

Misner, C. W., Thorne, K. S., and Wheeler, J. A., 1973, *Gravitation* (W. H. Freeman, New York, 1973).

Sagan, H., 1993, *Introduction to the Calculus of Variations* (Dover Publications, Mineola, NY, 1993).

Sakurai, J. J., 1964, *Invariance Principles and Elementary Particles* (Princeton University Press, Princeton, NJ, 1964).

Streater, R. F. and Wightman, A. S., 2000, *PCT, Spin, Statistics, and All That* (Princeton University Press, Princeton, NJ 2000).

Weinberg, S., 1972, *Gravitation and Cosmology* (John Wiley and Sons, New York, 1972).

Weinberg, S., 1995, *The Quantum Theory of Fields Volume I* (Cambridge University Press, New York, 1995).

Weyl, H., 1950, *Space, Time, Matter* (Dover, New York, 1950).

Weyl, H., (Tr. S. Pollard et al), 1987, *The Continuum* (Dover Publications, New York, 1987).

INDEX

About the Author

Stephen Blaha is a well known Physicist and Man of Letters with interests in Science, Society and civilization, the Arts, and Technology. He had an Alfred P. Sloan Foundation scholarship in college. He received his Ph.D. in Physics from Rockefeller University. He has served on the faculties of several major universities. He was also a Member of the Technical Staff at Bell Laboratories, a manager at the Boston Globe Newspaper, a Director at Wang Laboratories, and President of Blaha Software Inc and of Janus Associates Inc. (NH).

Among other achievements he was a co-discoverer of the "r potential" for heavy quark binding developing the first (and still the only demonstrable) non-abelian gauge theory with an "r" potential; first suggested the existence of topological structures in superfluid He-3; first proposed Yang-Mills theories would appear in condensed matter phenomena with non-scalar order parameters; first developed a grammar-based formalism for quantum computers and applied it to elementary particle theories; first developed a new form of quantum field theory without divergences (thus solving a major 60 year old problem that enabled a unified theory of the Standard Model and Quantum Gravity without divergences to be developed); first developed a formulation of complex General Relativity based on analytic continuation from real space-time; first developed a generalized non-homogeneous Robertson-Walker metric that enabled a quantum theory of the Big Bang to be developed without singularities at $t = 0$; first generalized Cauchy's theorem and Gauss' theorem to complex, curved multi-dimensional spaces; received Honorable Mention in the Gravity Research Foundation Essay Competition in 1978; first developed a physically acceptable theory of faster-than-light particles; first derived a composition of extrema method in the Calculus of Variations; first quantitatively suggested that inflationary periods in the history of the universe were not needed; first proved Gödel's Theorem implies Nature must be quantum; provided a new alternative to the Higgs Mechanism, and Higgs particles, to generate masses; first showed how to resolve logical paradoxes including Gödel's Undecidability Theorem by developing Operator Logic and Quantum Operator Logic; first developed a quantitative harmonic oscillator-like model of the life cycle, and interactions, of civilizations; first showed how equations describing superorganisms also apply to civilizations. A recent book shows his theory applies successfully to the past 14 years of history and to *new* archaeological data on Andean and Mayan civilizations as well as Early Anatolian and Egyptian civilizations.

He first developed an axiomatic derivation of the forms of The Standard Model from geometry – space-time properties – The Extended Standard Model. It has a Dark Matter sector that approximates the ElectroWeak sector with Dark doublets and Dark gauge interactions. It also uses quantum coordinates to remove infinities that crop up in most interacting quantum field theories and additionally to remove the infinities that appear in the Big Bang and generate an inflationary growth of the universe. The Extended Standard Model has an ultra-high energy GUT (Grand Unified Theory) limit with a $U(4)\otimes U(4)\otimes U(4)$ symmetry leading to 192 fundamental fermions stacked in four layers; and can be united with gravitation to form a Theory of Everything. Matter is shown to be 83.33% Dark in agreement with cosmological estimates. The color gluon sector lagrangian is shown to be compatible with the accepted Cornell Charmonium potential. A new boson quantum field theory was created to support the interpretation of negative energy states as classical boson fields in a manner analogous to Dirac's negative energy fermion sea theory. He has also shown how to transform coupling constants into to Higgsian vacuum expectation values with the remarkable result that the re-worked values of all coupling constants (including gravity) are comparable in value. He has also shown how Higgs particles originate in the transformation of complex-valued SU(2)⊗U(1) gsuge fields to real-valued gauge fields.

Recently he has developed the Strong Interaction and Gravitation subsector into a unified sub-theory that unifies color confinement and MOND – the galactic scale modification of gravity observed in Nature.

Blaha has had a major impact on a succession of elementary particle theories: his Ph.D. thesis (1970), and papers, showed that quantum field theory calculations to all orders in ladder approximations could not give scaling deep inelastic electron-nucleon scattering. He later showed the eigenvalue equation for the fine structure constant α in Johnson-Baker-Willey QED had a zero at $\alpha = 1$ not 1/137 by solving the Schwinger-Dyson equations to all orders in an approximation that agreed with exact results to 4th order in α thus ending interest in this theory. In 1979 at Prof. Ken Johnson's (MIT) suggestion he calculated the proton-neutron mass difference in the MIT bag model and found the result had the wrong sign reducing interest in the bag model. These results all appear in Physical Review papers. In the 2000's he repeatedly pointed out the shortcomings of SuperString theory and showed that The Standard Model's form could be derived from space-time geometry by an extension of Lorentz transformations to faster than light transformations. This deeper space-time basis greatly increases the possibility that it is part of THE fundamental theory.

In graduate school (1965-71) he wrote substantial papers in elementary particles and group theory: The Inelastic E- P Structure Functions in a Gluon Model. Phys. Lett. B40:501-502,1972; Deep-Inelastic E-P Structure Functions In A Ladder Model With Spin 1/2 Nucleons, Phys.Rev. D3:510-523,1971; Continuum Contributions To The Pion Radius, Phys. Rev. 178:2167-2169,1969; Character Analysis of U(N) and SU(N), J. Math. Phys. 10, 2156 (1969); and The Calculation of the Irreducible Characters of the Symmetric Group in Terms of the Compound Characters, (Published as Blaha's Lemma in D. E. Knuth's book: *The Art of Computer Programming Vols. 1 – 4*).

In the early 1980's Blaha was also a pioneer in the development of UNIX for financial, scientific and Internet applications: benchmarked UNIX versions showing that block size was critical for UNIX performance, developing financial modeling software, starting database benchmarking comparison studies, developing Internet-like UNIX networking (1982) and developing a hybrid shell programming technique (1982) that was a precursor to the PERL programming language. He was also the manager of the AT&T ten-year future products development database. His work helped lead to commercial UNIX on computers such as Sun Micros, IBM AIX minis, and Apple computers.

In the 1980's he pioneered the development of PC Desktop Publishing on laser printers. and was nominated for three "Awards for Technical Excellence" in 1987 by PC Magazine for PC software products that he designed and developed.

Recently he has developed a theory of Megaverses – actual universes of which our universe is one – with quantum particle-like properties based on the Wheeler-DeWitt equation of Quantum Gravity. He has developed a theory of a baryonic force, which had been conjectured many years ago, and estimated the strength of the force based on discrepancies in measurements of the gravitational constant G. This force, operative in 15-dimensinal space, can be used to escape from our universe in "uniships" which are the equivalent of the faster-than-light starships proposed in the author's earlier books. Thus travel to other universes, as well as to other stars is possible.

Blaha also considered the complexified Wheeler-DeWitt equation and showed that its limitation to real-valued coordinates and metrics generated a Cosmological Constant in the Einstein equations.

The author has also recently written a series of books on the serious problems of the United States and their solution as well as a book on the decline of Mankind that will follow from current social and genetic trends in Mankind.

In the past twelve years Dr. Blaha has written over 40 books on a wide range of topics. Some recent major works are: *From Asynchronous Logic to The Standard Model to Superflight to the Stars, All the Universe!, SuperCivilizations: Civilizations as Superorganisms, America's Future: an Islamic Surge, ISIS, al Qaeda, World Epidemics, Ukraine, Russia-China Pact, US Leadership Crisis,The Rises and Falls of Man – Destiny – 3000 AD: New Support for a Superorganism MACRO-THEORY of CIVILIZATIONS From CURRENT WORLD TRENDS and NEW Peruvian, Pre-Mayan, Mayan, Anatolian, and Early Egyptian Data, with a Projection to 3000 AD,* and *Mankind in Decline: Genetic Disasters, Human-Animal Hybrids, Overpopulation, Pollution, Global Warming, Food and Water Shortages, Desertification, Poverty, Rising Violence, Genocide, Epidemics, Wars, Leadership Failure.*

He has taught approximately 4,000 students in undergraduate, graduate, and postgraduate corporate education courses primarily in major universities, and large companies and government agencies.

The above paragraphs summarize much of his work over the past fifty years. This work is fully documented. He continues to engage in research and writing at Blaha Research.

www.ingramcontent.com/pod-product-compliance
Lightning Source LLC
Chambersburg PA
CBHW082007190326
41458CB00010B/3109